Green Energy and Technology

For further volumes:
http://www.springer.com/series/8059

Maria Kordjamshidi

House Rating Schemes

From Energy to Comfort Base

Maria Kordjamshidi
Ilam University
Faculty of Engineering
Pajoohesh
69315-516 Ilam
Iran
m_kordjamshidi@yahoo.com

ISSN 1865-3529 e-ISSN 1865-3537
ISBN 978-3-642-15789-9 e-ISBN 978-3-642-15790-5
DOI 10.1007/978-3-642-15790-5
Springer Heidelberg Dordrecht London New York

Library of Congress Control Number: 2010937671

© Springer-Verlag Berlin Heidelberg 2011

This work is subject to copyright. All rights are reserved, whether the whole or part of the material is concerned, specifically the rights of translation, reprinting, reuse of illustrations, recitation, broadcasting, reproduction on microfilm or in any other way, and storage in data banks. Duplication of this publication or parts thereof is permitted only under the provisions of the German Copyright Law of September 9, 1965, in its current version, and permission for use must always be obtained from Springer. Violations are liable to prosecution under the German Copyright Law.

The use of general descriptive names, registered names, trademarks, etc. in this publication does not imply, even in the absence of a specific statement, that such names are exempt from the relevant protective laws and regulations and therefore free for general use.

Cover design: Integra Software Services Pvt. Ltd., Pondicherry

Printed on acid-free paper

Springer is part of Springer Science+Business Media (www.springer.com)

Contents

1	**Introduction**		1
	1.1 Why House Energy Ratings Accomplished		1
	1.2 The Position of Passive Architecture in Current HERS		3
	1.3 A Conflict in HERS		3
	1.4 Book Outline – How to Read the Book		4
	References		5
2	**House Rating Schemes**		7
	2.1 House Energy Rating Schemes (HERS)		7
		2.1.1 The United States of America	8
		2.1.2 Canada	9
		2.1.3 Europe	10
		2.1.4 Australia	13
	2.2 Rating Methodologies for Buildings		14
		2.2.1 Building Rating Features	17
	2.3 Energy as the Main Parameter for Rating Buildings		18
	2.4 Issues Related to Building Energy Rating Schemes		19
		2.4.1 Rating and Achievement of Sustainability	19
		2.4.2 Rating Free Running Buildings	19
		2.4.3 Rating Index	20
		2.4.4 Occupancy Scenarios	21
		2.4.5 Accuracy of HERS	23
	2.5 Need for a New Index for Assessing Building Energy Efficiency		24
	2.6 Summary		25
	References		26
3	**Thermal Comfort**		31
	3.1 Thermal Comfort		31
		3.1.1 Definition of Thermal Comfort	32
		3.1.2 Human Comfort and Variables Affecting Thermal Comfort	32
		3.1.3 Thermal Comfort Models and Standards	34

		3.1.4	Applicability of the Thermal Comfort Index for Naturally Ventilated Buildings	36
		3.1.5	Adaptive Thermal Comfort Models for Naturally Ventilated/Free Running Buildings	39
		3.1.6	Acceptable Thermal Conditions in Free Running Buildings Based on the ASHRAE Standard	41
		3.1.7	Applicability of the Adaptive Comfort Model for Free Running Residential Buildings	41
	3.2	Evaluation of a Residential Building's Thermal Performance on the Basis of Thermal Comfort		43
		3.2.1	Computing Degree Hours for Free Running Houses	44
	3.3	Indicators to Measure the Thermal Performance of Houses for Rating Purposes		45
		3.3.1	Conditioned Mode	45
		3.3.2	Free Running Mode	45
		3.3.3	How an Indicator Points to Building Efficiency	46
	3.4	Summary		47
	References			47
4	**Modelling Efficient Building Design: Efficiency for Low Energy or No Energy?**			**53**
	4.1	Building Performance Evaluation		53
		4.1.1	Building Simulation Programs	55
		4.1.2	Criteria for Modeling the Thermal Performance of Buildings in Two Different Operation Modes	56
		4.1.3	Effective Parameters for Improving Buildings Thermal Performance	74
	4.2	Parametric Sensitivity Analysis of Thermal Performances of Buildings: A Comparative Analysis		77
		4.2.1	What Is Sensitivity Analysis?	77
		4.2.2	Thermal Performances of Dwellings in the Sydney Climate	77
		4.2.3	Summary of Thermal Performance Analysis	100
	4.3	Relationship Between Thermal Performance of Buildings on the Basis of Energy and Thermal Comfort		102
		4.3.1	Correlation Coefficient	103
		4.3.2	Multivariate Regression Analysis	108
	4.4	Conclusion		111
	References			113
5	**Assembling a House Energy Ratings (HER) and House Free Running Ratings (HFR) Scheme**			**117**
	5.1	Rating Building Thermal Performance		117
		5.1.1	How Should Building Thermal Performance Bands Be Defined for Rating?	118

	5.2 The Combination of Two Rating Systems	122
	5.3 How the New Combined System Evaluates Efficiency	125
	5.4 Reliability of the New Rating System	126
	5.5 Conclusion	128
	References	129
6	**Appendix**	131
	6.1 The Effect of House Envelope Parameters on the Seasonal Performance of Houses in Different Operation Modes	131
Index		141

List of Acronyms

HRS	House Rating Scheme
HERS	House Energy Rating Scheme
HFRS	House Free Running Rating Scheme
WCED	World Commission on Environment and Development
IEA	International Energy Agency
OECD	Organization for Economic Co-operation and Development
GHG	Green House Gas
MEC	Model Energy Code
BREDEM	Building Research Establishment Domestic Energy Model
SAP	Standard Assessment Procedure
NHER	National Home Energy Rating
AECB	Association for Environment Conscious Buildings
OEE	Office of Energy Efficiency
MNECH	Model National Energy Code of Canada
GMI	Government Metrics International
FSDR	Five Star Design Rating
CSIRO	Commonwealth Scientific and Industrial Research Organisation
BEP	Billed Energy Protocol
MEP	Monitored Energy Protocol
EEM	Energy-Efficient Mortgages
ITC	Index of Thermal Charge
GBA	Government Buildings Agency
HVAC	Heating, Ventilation and Air Conditioning Systems
ACS	Adoptive Comfort Standard
DDH	Degree Discomfort Hours
HW	Heavyweight
LW	Lightweight
SS	Single Storey
DS	Double Storey

Chapter 1
Introduction

1.1 Why House Energy Ratings Accomplished

During the mid to late 1970s the energy crisis, the increase in greenhouse gas emissions, and global warming concerns became important international issues. The importance of these issues for the future of humanity has led to international efforts in sustainable development. The "Brundtland Report" (1987) entitled "Our Common Future" showed that economic growth at the world's current rate was not sustainable on ecological grounds. The report saw potential climate change as an issue that threatened sustainable development, and recommended urgent action to increase energy efficiency. International concern about greenhouse gas emissions resulted in the Kyoto Protocol, which shared the 1992 United Nations Convention's objective for reducing greenhouse gas emissions by at least 5% from 1990 levels in the commitment period 2008–2012, with 168 countries, ratifying the protocol (United Nation, 1998).

However, according to the IEA (International Energy Agency, 2009), world energy consumption and carbon dioxide emissions have been increasing significantly. At the same time the International Energy Outlook report (Energy Information Administration, 2009) predicts "strong growth" for world-wide energy demand until 2030. This is shown in Fig. 1.1.

The impact of buildings on the environment is an important component in the consideration of sustainability. Buildings not only consume natural resources such as energy and raw materials, but also produce harmful atmospheric emissions. It is said that buildings consume one third of the world's resources (Atkinson, 2006), which includes approximately one third of primary energy supply (Hong et al., 2000). That means that buildings are an important contributor to global warming as well.

The residential building sector is seen to be a significant contributor to energy consumption and subsequently to greenhouse gas emissions. Residential buildings have been found to be responsible for emitting about 15% of greenhouse gas emissions in OECD[1] countries such as Australia (Harrington et al., 1999), US (United

---

[1] Organization for Economic Co-operation and Development (OECD), consisting of 30 countries.

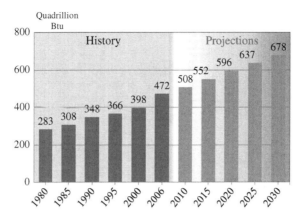

Fig. 1.1 World market energy consumption, 1980–2030 (Source: Energy Information Administration (EIA), International Energy Annual (June–December 2008), World Energy and Economic Outlook (2009))

Nations, 2004) and UK (Office of the UK Deputy Prime Minister, 2004). Meanwhile the IEA (2007) has projected that residential energy end-use will rise by an average of 1.7% per year. Thus in most International Energy Agency countries, the residential sector has been the focus of more energy related policies than any other sector.

In response to the call for reducing energy demand in the building sector, House Energy Rating Schemes (HERS) have been developed in order to promote energy efficient design. These schemes offer a means for comparing the energy efficiency of different homes by generally providing a standardized evaluation of a home's existing energy efficiency, expected energy use cost, and its potential for improvement. They differ in the range of energy end use categories covered, but commonly the basis of most programs has been the normalized energy requirement for space heating and cooling, and sometimes water heating. However, relying on the control of energy consumption is not the only way to achieve energy efficiency in architectural design.

The main objective of HERS is to reduce energy consumption and greenhouse gas emissions. They mainly operate through the calculation of the predicted energy requirements of buildings in order to enable energy conservation and energy efficient building design. HERS have been created to make it possible for energy efficiency to become an explicit component in home evaluation and thermal performance assessment, and thus in the purchasing decision process. Vine et al. (1988), Turrent and Mainwaring (1990) and Ballinger and Cassell (1994) have all argued that HERS are one of the most successful methods for improving residential energy efficiency in developed countries. This claim is supported in many nations which have created a link between financing (through mortgages) and HERS, to support energy efficient design (Farhar et al., 1996).

Clearly the most effective way of achieving efficient architecture, and thus the main objective of HERS, would be the promotion of passive architectural design, with value placed on the efficiency of house design in "free running operation".[2] However, HERS are limited in their effectiveness, since they generally ignore the significance of passive architectural design as a means of ensuring energy efficiency (Soebarto, 2000; Williamson, 2000).

1.2 The Position of Passive Architecture in Current HERS

The current House Energy Rating Schemes discriminate against free running houses[3]; and therefore may be said to discourage architects and designers from promoting passive architecture buildings. This is a result of the fact that a systematic method for the evaluation of the free running performance of houses is missing in current house rating schemes. While the main objective of these schemes has been a reduction in energy consumption and greenhouse gas emissions in the building industry, they have been developed on the basis of predicted energy requirements involving the use of active heating and cooling. They do not deal at all with free running buildings designed to largely *avoid* heating and cooling. A building obtains a higher score through such a scheme if the predicted active energy use is low compared to the defined standard reference for the system. Logically, however, the highest score should actually be attached to a passive architectural design with no need for artificial energy for space heating and cooling. In fact, the result is that under the current house rating schemes the passive climate control features of houses may be sacrificed to pay for air conditioners. This becomes a significant issue in the moderate climates of some regions, such as in Australia, in which passive architectural design could be said to be the most suitable response for achieving the objectives of HERS, and should therefore be taken into consideration in building regulatory frameworks.

1.3 A Conflict in HERS

There is a growing demand for space heating and cooling (EIA, 2009) as people demand a higher level of indoor comfort in modern society, using air-conditioning,[4] and this is a problem which is unlikely to be solved under present conditions because

---

[2] In this book the definition of *free running* that is used is: the state of a building that is naturally ventilated and does not use any mechanical equipment to maintain or improve its indoor thermal condition. In contrast, those buildings that are provided with an energy supply applied to heat/cool air or surfaces to maintain indoor conditions within a defined comfort zone are referred to as operating in *conditioned mode*.

[3] This issue is discussed in detail in the next chapter (See Sect. 2.5).

[4] See the EIA web site *comparison with other projections,* at: http://www.eia.doe.gov/oiaf/aeo/pdf/forecast.pdf

of the dependency of current house rating schemes on calculating energy consumption. This situation is exacerbated by the fact that while authorities have been trying to encourage the application of efficient building design, with higher ratings reflecting decreased energy demand, people tend to think that the higher score means a higher level of comfort. This tendency results in a "take back effect", which occurs when people with more efficient homes actually use more energy than expected because they are less cautious about basic efficiency measures such as thermostat settings (Stein, 1997b). It has been noted that despite efforts to improve energy performance, currently "houses do not perform optimally" (Wray et al., 2000); in other words, the thermal performance of houses when occupied is not such as has been expected or intended.

Research has demonstrated a number of shortcomings in the current rating schemes that mean that they have been unable to reach the desired objectives of sustainability (Kordjamshidi, 2008; Soebarto, 2000; Stein, 1997a; Stein and Meier, 2000; Williamson, 2000). These shortcomings, namely the inaccuracy of ratings, unrealistic standardised occupancy scenarios, and the unreliability of a normalized index for evaluating the thermal performance of buildings, will be discussed in more detail in this book.

1.4 Book Outline – How to Read the Book

The main objective of this book is to introduce a new framework for a House Rating Scheme by which the efficiency of the architectural design of all houses can be evaluated without unrealistically compromising the value of any particular design. This book provides information about developments in the field of Building Energy Ratings, concentrating on House Energy Rating Schemes. In reviewing current House Rating Schemes developed in different countries, the book describes how these schemes assess the thermal performance of a house, and challenges the way that these schemes assess building energy efficiency, and their inability to evaluate free running buildings. It deals with various approaches and methods for rating buildings on the basis of different indexes, with implications for both energy efficiency and thermal comfort. It also guides readers through a computer simulation program for developing a rating system that evaluates and ranks building energy efficiency.

The book is made up of five chapters:

> Chapter 2 describes the current House Energy Rating Schemes (HERS) to specifically identify shortcomings in the current HERS. It points to the necessity of revising current building rating systems towards a thermal comfort base rating scheme in order to deal with rating free running houses.
> Chapter 3 describes the theoretical aspects of thermal comfort in buildings to establish how thermal comfort and its measurable parameters can become

the basis for a rating scheme. In this chapter a specific indicator of thermal comfort is defined for measuring the thermal performance of buildings.

Chapter 4 deals with building performance evaluation – methods and criteria for assessing and ranking the thermal performances of dwellings in different operation modes. This chapter, using statistical analysis, identifies significant differences between building thermal performance in free running and conditioned operation modes.

Chapter 5 presents a method for rating buildings thermal performance in both free running and conditioned operation modes. This chapter presents a simplified framework for a free running rating scheme for dwellings, and then proposes a new framework for HRS. The general utility of the proposed rating framework is then tested and described.

References

Atkinson, M.: The impact of building on the environment – what's needed to change the status quo? Green Building Council Australia. www.gbcaus.org (2006). Accessed 30 Aug 2006

Ballinger, J.A., Cassell, D.: Solar efficient housing and NatHERS: an important marketing tool. Proceedings of the Annual Conference of the Australian and New Zealand Solar Energy Society, Sydney, pp. 320–326 (1994)

Brundtland, G.H.: Our Common Future – Report of the World Commission on Environment and Development. Oxford University Press, Oxford (1987)

Energy Information Administration (EIA): World Energy Demand and Economic Outlook. http://www.eia.doe.gov/oiaf/ieo/pdf/world.pdf (2009). Accessed 10 Feb 2010

Farhar, B.C., Collins, N.E., Walsh, R.W.: Linking Home Energy Rating Systems with Energy Efficiency Financing: Progress on National and State Programs (No. NREL/TP-460-21322): National Renewable Energy Laboratory (1996)

Harrington, L., Foster, R., Wilkendfeld, G., Treloar, G.J., Lee, T., Ellis, M.: Baseline Study of Greenhouse Gas Emissions From the Australian Residential Building Sector to 2010. Canberra: Australian Greenhouse Office (1999)

Hong, T., Chou, S.K., Bong, T.Y.: Building simulation: an overview of developments and information sources. Build. Environ. 35(4), 347–361 (2000)

United Nations: Framework Convention on Climate Change (Report on the in-depth review of the third national communication of the United States of America. No. FCCC/IDR.3/USA). United Nations, New York, NY (2004)

International Energy Agency: Energy projection. http://www.iea.org/Textbase/subjectqueries/keyresult.asp?KEYWORD_ID=4107 (2009). Accessed 12 Dec 2009

Kordjamshidi, M.: In Australia energy based rating tools appear to have failed to deliver the policy outcomes for sustainable development, Proceeding of 12th Passive house conference, 11–12, 2008, Nuremberg, Germany (2008)

Office of the UK Deputy Prime Minister: Government Moves Ahead with Developing New Code for Sustainable Buildings. http://www.odpm.gov.uk/pns/DisplayPN.cgi?pn_id=2004_0181 (2004). Accessed 27 July 2007

Soebarto, V.I.: A Low-Energy House and a Low Rating: What is the Problem, Proceedings of the 34th Conference of the Australia and New Zealand Architectural Science Association, Adelaide, South Australia, pp. 111–118 (2000)

Stein, J.R.: Accuracy of Home Energy Rating Systems (No. 40394). US: Lawrence Berkeley National Laboratory (1997a)

Stein, J.R.: Home Energy Rating Systems: Actual Usage May Vary. Home Energy Magazine Online. http://hem.dis.anl.gov/eehem/97/970910.html (1997b, Sep/Oct). Accessed 11 June 2008

Stein, J.R., Meier, A.: Accuracy of home energy rating systems. Energy. **25**(4), 339–354 (2000)

Turrent, D., Mainwaring, J.: Saving energy on the rates. RIBA J. 85–86 (1990, Sep)

International Energy Agency: Energy Projection. http://www.worldenergyoutlook.org/2007.asp (2007). Accessed 10 Oct 2007

United Nations: Kyoto Protocol to the United Nations Framework Convention on Climate Change. United Nations. http://unfccc.int/kyoto_protocol/items/2830.php (1998). Accessed 15 Jan 2007

Vine, E., Barnes, B.K., Ritschard, R.: Implementing home energy rating systems. Energy. **13**(5), 401–411 (1988)

Williamson, T.J.: A critical review of home energy rating in Australia. Proceedings of the 34th Conference of the Australia and New Zealand Architectural Science Association, Adelaide, South Australia, pp. 101–109 (2000)

Wray, C.P., Piette, M.A., Sherman, M.H., Levinson, R.M., Matson, N.E., Driscoll, D.A. et al.: Residential Commissioning: A Review of Related Literature (No. LBNL_44535). US: Lawrence Berkeley National Laboratory (2000)

Chapter 2
House Rating Schemes

This chapter presents selected House Energy Rating Systems in diverse contexts and explores the different aspects of a House Energy Rating Scheme (HERS). It demonstrates that there are inadequacies in the current rating schemes which this book attempts to address.

2.1 House Energy Rating Schemes (HERS)

The *energy* rating of a house is a standard measure that allows the energy efficiency of new or existing houses to be evaluated, in order that dwellings may be compared. The comparison is commonly performed on the basis of the energy requirements for the heating and cooling of indoor spaces. Some of the HERS include all energy requirements, such as energy for water heating, washing machines and cooking.

Energy is not the only criterion for house evaluation in all rating schemes. Criteria are determined on the basis of the purpose of the rating. Other criteria that have been used as important parameters in building evaluation systems are the production of greenhouse gas (GHG) emissions, indoor environment quality, cost efficiency and thermal comfort.

The energy rating of a residential building can provide detailed information on the energy consumption and the relative energy efficiency of the building. It is performed through standard measurements carried out under specific regulations and experimental procedures by specialists (Santamouris, 2005). Overall, HERS can facilitate informed decision-making for all stakeholders, as well as home-buyers considering mortgages. The main impetus behind most of the rating systems has been to inform consumers about the relative energy efficiency of homes, in order to encourage home-owners to use this information in making their purchasing decisions (SRC, 1991).

HERS are found in a variety of forms:

- prescriptive
- calculation-based
- performance based

All of those evaluate building performance within the scope of a program that has been developed by the authorities of a country to promote efficiency in building design. *Prescriptive* schemes provide minimum standards for the materials, equipment and methods of efficient design and construction that must be met to qualify for an energy efficiency rating. *Calculation* based ratings employ computer based models to predict a building's performance relative to that required in order to qualify for a rating under the program. *Performance* based ratings utilize actual building energy consumption data to evaluate building energy efficiency, which is then compared with the required standards of the program.

Prescriptive and calculation schemes are predominant, whereas performance based rating schemes are very rare because of the time-consuming nature of the system, which requires an extensive effort. Performance based schemes are also not applicable to new buildings because of their limited value as a tool for predicting performance and encouraging improvements prior to construction.

Rating schemes are generally associated with either *certification* or *labelling*. The former refers to the evaluation of building performance at the design stage, while labelling assesses in-use performance of the building when it is compared with other similar buildings.

The schemes vary in practice, from simply a paper-based check-list, to full thermal simulations. A good example of a paper-based check-list is the Model Energy Code (MEC) (Andersen et al., 2004), which was developed for the Department of Energy Building Standards and Guidelines Program in the United States. MEC focuses on the insulation of the envelope and windows of a building, the cooling and heating system, the water heating system, and air leakage. Most of these rating schemes use a grading scale to score buildings. One hundred point scales and star rating systems are common, while some use either a pass/fail system, or simply classify by terms such as bronze, silver, or gold. MEC is a simple pass or fail scheme (US Department of Energy, 1995).

Generally, all developed rating schemes around the world appear to be similar in their objectives, but different in programming and details. A general review of developed HERS has shown that these schemes are particularly widely implemented in the USA. The following section reviews HERS programs that have been actively implemented in the United States, Europe, Canada and Australia.

2.1.1 The United States of America

Energy rating schemes have been used in the USA since the 1980s (Santamouris, 2005). Over the past years a range of rating schemes has been implemented by the different states, cities, utilities and vendors. There are a variety of efficiency certification programs and numerous tools for analysing building performance.

Among the various schemes, the Energy Rated Homes of America is predominant, as it is currently operating in more than 18 states, with other schemes in continuous development in the other states. This scheme uses a 100-point scale of efficiency, divided into ten categories of stars (from one star, one star plus, to five

2.1 House Energy Rating Schemes (HERS)

stars plus). A higher star represents a house with better energy efficiency. The energy efficiency rating in this system expresses the predicted energy consumption, which is represented in the form of normalised annual energy consumption. The dependency of this rating system on a calculation of the amount of energy consumed means that the use of efficient appliances results in a more favourable rating than that for an efficient architect designed house, whereas arguably a free running house should have priority for reducing energy consumption.

Numerous software programs have been developed to foster increased energy efficiency in the building sector. In North America alone there exist about a hundred building energy tools serving a diversity of users (Mills, 2004). Many of these are applied to rate buildings, such as AkWarm, Building Greenhouse Rating, LEED, CHEERS, RECA 2000, Kansas, HOT 2000, Ohio, REM/Rate, TRET, Energy Gauge USA, T. A. P, BESTTEST, HEED, Colorado and E-Star.[1]

The main objectives of the Home Energy Rating Schemes implemented in the USA are: affordability (a higher quality and more comfortable home for less money), qualifying for a more favourable mortgage loan, and environmental protection (through optimizing residential and commercial energy and indoor environmental performance). The association of home energy rating systems, with a scheme called Energy Efficiency Mortgages, brought about the penetration of this rating system into the residential market (Santamouris, 2005). The mortgage industry uses existing energy audits to make loans for energy improvements (Barbara, 2000).

2.1.2 Canada

The Office of Energy Efficiency (OEE) has developed and promoted a wide range of programs in Canada. These are aimed at improving energy efficiency in the energy sector of the Canadian economy, at conserving energy resources, aiding financial savings and reducing greenhouse gas emissions.

Home energy rating systems for houses in Canada, which began in 1997, were based on the report "Efficiency of Natural Resources Canada" (NRCan). There are two national energy rating programs for residential buildings, named Ener-Guide for Houses (EGH) and Ener-Guide for New Houses (EGNH). These governmental programs use HOT2XP and HOT 2000 as their rating tools. The tools are programmed to make a comparison for rating purposes of each house, with reference to houses of a similar size in a similar climatic region. To factor out the influence of occupants on energy consumption, standard operating conditions are used in calculating the rating. The energy rating assessment begins with a site evaluation, using a blower

---

[1] More details about software programs can be found in the web-based references given by the US Department of Energy, 2009: *Building energy software tools directory,* <Energy Rated Homes of America US Department of Energy. http://apps1.eere.energy.gov/buildings/tools_directory/subjects_sub.cfm.

door test to measure the rate of air leakage in homes. The space heating and cooling systems and domestic hot water supply, appliance usage, and mechanical systems are analysed to produce an energy efficiency rating based on the home's annual energy consumption, on a rating ranging from 0 to 100 (Allen, 1999). The lower rating on the scale indicates high leakage, no insulation, high-energy consumption and therefore an uncomfortable home to live in.

Two standard bases for evaluating buildings are R-2000 standards[2] and the Model National Energy Code of Canada (MNECB).[3] To meet Canada's specifications Code, a house needs to be rated within the 80–85 range to comply with R-2000, or in the 70–75 range to comply with MNECH (Allen, 1999). The softwares used for analysing a building's performance are: HOT 2000, HOT 3000, HOT2XP, HOT2EC, EE4, GBtool and BILDTRAD. All can evaluate the energy performance of a building, but are unlikely to be applicable for a free running house evaluation.

One program used for the Canadian homes rating system is LEED. This rating system is an adaptation of the US Green Building Council's LEED Green Buildings Rating System, tailored particularly for Canadian climates, construction and regulations. This rating system measures the overall performance of a home in eight categories: Innovation and Design Process (ID), Location and Linkages (LL), Sustainable Sites (SS), Water Efficiency (WE), Energy and Atmosphere (EA), Material and Resources (MR), Indoor Environmental Quality (EQ), Awareness and Education (AE). The rating system works by requiring a minimum level of performance through prerequisites, and rewarding improved performance in each of the eight categories. The performance level is indicated by four grades: Certified, Silver, Gold and Platinum, based on the number of points gained (between 45 and 136 points) (Canada Green Building Council, 2009).

2.1.3 Europe

Following the energy crisis in the 1970s, preliminary steps for energy saving measurements in Europe occurred in Sweden. Since 1993 a "Specific Actions for Vigorous Energy Efficiency Directive" has been employed throughout the countries in the European Union (Cook et al., 1997). The aim has been to "certify" the energy efficiency of homes. Since the directive neither specifies the certification

---

[2] The R-2000 Standard is based on an energy consumption target for each house, and a series of technical requirements for ventilation, air tightness, insulation, choice of materials, water use and other factors (See: http://oee.nrcan.gc.ca/residential/personal/new-homes/r-2000/About-r-2000.cfm?attr=4).

[3] [MNECB] is intended to help in designing energy-efficient buildings. It sets out minimum requirements for the features of buildings that determine their energy efficiency, taking into account regional construction costs, regional heating fuel types, and costs and regional climatic differences. The MNECB has, in addition to sections on the building envelope and on water heating, detailed information on lighting, HVAC systems and electrical power, which can offer major energy savings (See: http://irc.nrc-cnrc.gc.ca/pubs/codes/nrcc38731_e.html).

2.1 House Energy Rating Schemes (HERS)

procedure, nor identifies the kind of energy that should be assessed, the states were requested to prepare their own national methodologies (Santamouris, 2005), and each member country has produced a different interpretation of the term "certification". The European Energy Commission then put forward a proposal for a new specific directive on the energy rating of buildings (based on "Energy Performance of Building Directive (EPBD) 2002/91/EC 16"). The EU adopted EPBD, which provided a common methodology for calculating the energy performance of buildings, and set minimum efficiency standards for residential and commercial buildings. The directive then introduced an energy performance certificate to promote greater public awareness. However, there are still no standards for the energy performance of existing buildings in the EU.

A review of the energy ratings of dwellings in the European Union by Miguez et al. (2006) describes the various rating systems in EU countries. Current rating systems, based on several regulations, all aim to save energy and reduce greenhouse gas emissions. These rating systems assess a building as to whether it complies with regulations. A range of techniques has been developed for such building assessment, and all are based on an experimental protocol for collecting energy data and theoretical algorithms to normalize total energy consumption for classifying buildings. Total energy consumption results from heating, hot water supply and lighting. Because of high heating energy requirements, all the member states in the EU have introduced compulsory maximum levels for coefficients of heat transmission in new buildings. The cold climate in these countries demands more insulation generally, meaning lower energy losses and GHG emissions.

Although the preliminary steps for energy saving and efficient energy use in the building sector were taken in Sweden, this nation still has no official energy rating system for buildings. However they do have stringent regulations. Among different rating systems in the EU, Denmark's is known as the system which provides full energy rating, in the sense of awarding a graded score to buildings. The ratings developed in the UK and Denmark are discussed in more detail below, as they are the two pioneering rating systems in the EU.

2.1.3.1 United Kingdom

The oldest HERS exists in the United Kingdom. It mainly aims to decrease energy consumption and GHG emissions. Two house energy-rating schemes are currently operating in the UK. The National Home Energy Rating scheme (Hasson et al., 2000) was developed and implemented by the National Energy Foundation, an independent charitable trust (Turrent and Mainwaring, 1990). This scheme measures the thermal efficiency of dwellings in terms of energy running costs on a scale of 0–10. The rating procedure is carried out through the use of a computer program based on the Building Research Establishment Domestic Energy Model (BREDEM). This is used in different ways as the basis of the Standard Assessment Procedure (SAP), National Home Energy Rating (NHER) and CO_2 Dwelling Emission Rate (DER) (Energy Efficiency Partnership for Homes, 2006). In BREDEM the energy usage of a house is calculated on the basis of a description of its dimensions, insulation and heating system.

The Standard Assessment Procedure (SAP) has been developed by British planning authorities as the principal basis for labelling and house rating. It was drawn up to define the method of energy rating of residential buildings (Miguez et al., 2006; Richalet and Henderson, 1999). Energy rating is based on energy balance and cost for space and water heating per square meter of floor area, assuming average occupancy patterns. It includes details of the house, such as the heating system, thermal insulation, ventilation characteristics and the type of fuel used for heating, as factors affecting energy efficiency. Fuel costs and gas emissions are assessed, and on the basis of this individual suggestions for improvements are given. This rating does not consider lighting and domestic appliances in the process of calculating energy consumption, and it ignores the location of the building for the rating purpose. These omissions would appear to have a significant effect on the accuracy of the rating system, and to potentially discriminate against the value of a building design which might be suitable for a particular location and climate.

As there were doubts about its ability to achieve the target of energy saving and reduction of GHG emissions in the building sector, the SAP regulations were revised in 2001 (DEFRA, 2005). Nevertheless, as the basis of the methodology for improving the energy efficiency of buildings continues to be the calculation of energy consumption, it may well not be accurate in providing passive energy measurements, as demanded by Association for Environment Conscious Buildings (AECB, 2006), and is unlikely to grade passive architecture designs accurately.

NHER measures the energy efficiency of houses as a function of energy running costs per square meter. It calculates energy usage by taking into account the house details, including house location, design, construction, water heating system, cooking, lighting, ventilation and appliances. To calculate the rating, a standard occupancy scenario is assumed, in which the number of occupants is estimated from the house floor area and standard heating patterns. Thermostat settings and the period of occupation are also included as part of the standard. The actual occupancy data can be used to estimate the running costs, fuel use and emissions, but this will not alter the rating.

2.1.3.2 Denmark

As a pioneer in energy rating in the EU, Denmark started energy saving measurement in 1981. This country established a different type of energy audit, known as the "Act on the Promotion of Energy and Water Conservation in Buildings" (Energies-Cites, 2003; International Energy Agency, 2003). It comprises energy certificates for large and small buildings as well as for industrial buildings, and for CO_2 emissions in industry (Miguez et al., 2006).

The rating system is based on an energy inventory recorded by a qualified specialist. It includes three parts. The first part reports on water and energy consumption and CO_2 emissions per annum as compared with other similar buildings, on a rating scale from A1 to C5 (maximum to minimum efficiency). An energy plan is the

2.1 House Energy Rating Schemes (HERS) 13

second part of the system, through which ways for saving energy and water in buildings are proposed, with an estimation of the costs involved, and annual savings for each one. The final section of the rating provides information on the current state of the building in terms of its size, heating system and energy usage, and the cost of energy and heating.

This rating system appears to be sufficiently comprehensive for conditioned buildings but it is not able to deal with rating free running houses, owing to its dependency on the energy base.

2.1.4 Australia

House Energy Rating Schemes have also been introduced in Australia, with the same objectives as those in the other mentioned countries. The main objectives are to decrease residential energy consumption and greenhouse gas emissions, and to increase thermal comfort by encouraging improved building envelope design (Ballinger, 1998a).

Where the Australian climates differ from those of Europe and Canada, differences in the programming of HERS in Australia were expected. "It has been shown in many studies that passive solar design and energy conservation techniques are very cost-effective in Australia. Australian climates allow us to enjoy the outdoor generally throughout the year except on days of temperature extremes" (Ballinger, 1988, p 67). The moderate climate of some regions in Australia makes passive architectural design such as free running houses a good option, and most suitable for achieving the objectives of HERS. However, house ratings in Australia, as in other countries, are based on the prediction of energy requirements, and have not been modified to give more value to free running houses.

The Five Star Design Rating was the first energy-rating scheme, developed in Australia in the 1980s by the GMI Council of Australia. It was adopted for use in Victoria, New South Wales and South Australia. "Five Star Design Rating" (FSDR) is a form of certification available for dwelling buildings which comply with a number of requirements for energy efficient design. The design principles of a five star home under this system were based on the three basic elements of glass, mass and insulation (Ballinger, 1988). However, this system was not widely accepted by the building industry, because of its restrictive guidelines and its limitation to a single pass/fail rating.

During the 1990s, individual states in Australia attempted to develop their own House Energy Rating Schemes (HERS) to meet particular needs (Ballinger, 1991; Gellender, 1992; Wathen, 1992). Among the different schemes, the Victorian scheme, based on a computer program, was found to be the most effective; however, it was not flexible enough for all climates, particularly for warm humid climates such as in Queensland. It was therefore thought appropriate to develop a nationwide HERS.

The development of a nationwide House Energy Rating Scheme (HERS) was started in 1993 on behalf of the Australian and New Zealand Minerals and Energy

Council (Ballinger and Cassell, 1994; Szokolay, 1992b). The aim was to create a simple rating for energy efficiency for each dwelling throughout different climate zones and conditions in Australia. A graded five-star rating system was used to categorize the relative energy efficiency of dwellings, using a computer program based on the CHEETAH (engine), which was developed for the rating assessment (Ballinger and Cassell, 1994).

HERS predict the demand for the heating and cooling energy required to maintain conditions of thermal comfort inside a building, and rate the building's average energy consumption per square meter (MJ/m^2). Predictions are based on the extensive research and development embodied in CHEENATH, the core energy software model developed by CSIRO as suitable for Australian climates (Ballinger, 1998a). This engine, which is a significantly enhanced version of the CHEETAH engine, is the current basis of most modelling systems, such as NatHERS, FirstRate and Quick Rate, BERS, Q Rate and ACTHERS, which have been developed in different states. NatHERS and BERS simulate the operational energy usage in a home by running CHEENATH directly (with different user interfaces), while FirstRate, QRate, ACTHERS and Quick Rate are correlation programs, which do not carry out simulations.

AccuRate is the latest tool developed for HERS. It addresses some of the limitations in the NatHERS software and is now a replacement for NatHERS. More details regarding this are presented in Sect. 4.1.2.2.

At the time when NatHERS was created, it was assumed that this software would be developed in the future on the basis of comfort achieved without the use of heating and cooling (Ballinger, 1998b). However this project still remains incomplete, even in the latest developed tool for HERS.

2.2 Rating Methodologies for Buildings

Buildings present many characteristics that need to be taken into account for an appropriate evaluation and rating scheme (Roulet et al., 1999). Thus a wide range of rating methods has been developed.[4] Each method considers a number of parameters and criteria to assess buildings on a particular basis. These include perceived health, the provision of thermal, visual and acoustic comfort, indoor air quality, cost effectiveness, environmental impact and energy efficiency. However, energy efficiency is seen as the main parameter in almost all current building rating schemes, even in those which aggregate and evaluate buildings on the basis of a multi-criteria method.

The various methodologies developed to evaluate the energy efficiency of buildings are principally based on predicting energy consumption to assess a building, in order to certify the level of the building's performance (Santamouris M., 1995; Boland et al., 2003, Richalet et al., 2001; Santamouris, 2005; Santamouris and

---

[4] Some of these methods appear in a review by Kotsaki, K. and G. Sourys (2000).

2.2 Rating Methodologies for Buildings

Dascalaki, 2002; Santamouris et al., 2007; Zmeureanu et al., 1999). The method is the same whether the building is residential or contains office space.

A historical review by Fairey et al. (2000) of the national HERS methods used in the US describes the following four proposed methods for rating the energy efficiency of homes:

- the original method
- the equipment adjustment factor method
- the modified loads method
- the normalized modified loads method.

Each method was developed to overcome shortcomings in the previous method. In the *original method* the score of a home depended on the fraction that the total estimated purchased energy consumption of the house represented of that for a reference home.[5] The dependence of this method on the fuel type involved represents a "flaw" in the method. This "flaw" is due to the "floating" value of the reference house, whose value could change as the function of a selected fuel type, and consequently the score of a home could simply change. This problem was solved through the second method, by adding an *equipment adjustment factor*. However, the main issue with the second method was that "rating directly by energy consumption misrepresents the relative value of envelope efficiency measures with respect to equipment efficiency measures" (Fairey et al., 2000, p. 4). *The modified loads method* was then developed to avoid the above problem. In this method, building loads[6] were used instead of energy consumption, to establish the rating fraction used in the original method. Since the load on building end uses does not change as a function of fuel, the "floating" problem was also solved. However, the presence of a "fuel neutrality flaw" was a problem with this method, due to the fact that different fuel types may be discriminated against in marketing. A "normalized modified load method" was then proposed that reflects differences in potential equipment improvements.[7]

The first two methods rely on the calculation of energy consumption, while the last two refer to the amount of energy load. Although the second basis is more reliable, both bases have the shortcoming that they are unable to exactly predict actual energy consumption and energy load, because certain variables such as occupancy and the behaviour of occupants could change the results of the calculation.

Botsaris and Prebezanos (2004) introduced a method for the certification of the energy consumption of a building by recording its "energy behaviour". In this

---

[5] A home score in the original method was calculated from 100 to 20*(ER/EC), in which ER is the total purchased energy consumption for heating, cooling and hot water for the rated home, and EC is that for the reference home.

[6] Load in this method is defined as the amount of heating energy that must be added or removed from a building to satisfy a specified level of comfort in the building, and Energy Use is the amount of energy required by the equipment that satisfies the load.

[7] The mathematical process is described in Fairey et al. (2000).

method energy indices, such as the Index of Thermal Charge (ITC) or Index of Energy Disposition (Andersen et al., 2004), are employed to simulate the heat losses of the building, and the heat flow due to temperature difference between the indoor and the outdoor space. This work is based on an interpretation of the behaviour of the energy sources, such as the operation and cessation time of the sources. Cessation times can be predicted relatively reliably for office buildings with a clear occupation time. However, this method for residential buildings may not be accurate, owing to the variability of its occupants' activities. The method can, however, help to accurately predict the energy consumption of residential buildings if it is adapted to include *multiple occupancy scenarios*.

A review of the latest developments in the field of the-energy rating of dwellings, mainly in Europe, describes the theoretical and experimental techniques for energy characterization of buildings that have been employed (Santamouris, 2005; Miguez et al., 2006) and shows that all of the systems have been developed basically to predict the total energy demand of a building.

EUROCLASS is a recent method developed for the energy rating of buildings through the European SAVE program. It suggests a theoretical technique that comprises all specific energy uses and treats energy normalization in a new manner. It proposes a new framework based on the use of "the relative frequency distribution curves for the different end users of the energy" (Santamouris, 2005, P71). The variables which are determined to grade a building are "total supplied energy" (kWh/m^2) and "total delivered energy" (kWh/m^3). These variables can be obtained from two protocols: the Billed Energy Protocol (BEP) and the Monitored Energy Protocol (MEP). Each of these protocols provides useful information for carrying out a rating test of a building in a specific comparison scenario. EUROTARGET is the software developed within the frame of the EUROCLASS project to apply this proposed rating methodology for dwellings.

There are a number of studies that propose multi-criteria for a building assessment and rating scheme. These studies include a number of parameters to rank buildings, such as energy use for heating and cooling, indoor environment quality, cost, impact on the external environment and the life-cycle of the embodied energy of construction (Roulet et al., 2002; 2005; Soebarto and Williamson, 2001).

In the study by Soebarto and Williamson (2001) a methodology based on a weighting method was developed to assist the building design process and assess a building's environmental performance in accordance with multi-criteria assessment. This methodology converts the criteria into a two criteria problem by creating a weighted sum of benefits and costs for each solution. These two functions are normalized to reflect the average weighting value. An environmental performance assessment tool, ENE-RATE, was developed on the basis of this method to perform environmental ratings. Although this study accepts that thermal comfort in an unconditioned building should be considered as a criterion for building evaluation, it does not clarify any method of incorporating that criterion for that purpose.

Roulet et al. (2002) produced a multi-criteria ranking methodology to rate office buildings. The method employs fuzzy logic on a set of indices, each of which addresses a particular aspect of building performance in the two categories of energy and comfort. Using a principal components analysis, the energy and comfort parameters are combined in a single indicator that globally characterizes the performance of the building. Annual energy use for heating, cooling and lighting (kwh/m^2) and discomfort hours during winter and summer (h) are the criteria used to define this single indicator. The proposed criteria for indoor environment quality are: predicted percentage of dissatisfaction based on the Fanger comfort model,[8] outdoor airflow rate per person, and noise level in the working place. Each parameter is given a weight depending on the scale of values of the user of the method. This method would not appear to be successful in evaluating thermal comfort conditions in a naturally ventilated building, because the employed criteria are only applicable for conditioned buildings. The method can be adapted for use in any multi-criteria rating scheme.

Regardless of the function of a building, *normalised energy use* is seen as the most common method to evaluate the efficiency of a building in the conditioned operation mode (Chung et al., 2006). This method regards the building's size and annual energy use, divided by the conditioned floor area or by volume. There are shortcomings in this method which make it unrealistic for addressing the efficiency of an architectural design. This will be discussed later in Sect. 2.4.3.

2.2.1 Building Rating Features

Almost all of the rating schemes address the features of the building's envelope and the efficiency of equipment for cooling, heating indoor space, and hot water supply. Some of them include energy related fixed components such as washing machines, dishwashers, refrigerators, freezers and dryers. Current tools employed for rating systems have the capability of calculating heating, cooling, hot water, lighting, and appliance energy loads. Some of them also predict the energy cost of new and existing single and multifamily homes, on the basis of the prediction of total energy requirement, the type of fuel used, and the efficiency of appliances. Occupancy factors have usually been considered as a default or are standardized; however, a limited number of ratings tools are flexible enough to change the occupancy variables, such as the number of occupants and the hours of occupation.

There are many similarities between the different systems. They all use some combination of data collection and calculation to present information to building users about energy consumption. Their reliance on calculation is almost inevitable because of the highly disparate nature of buildings. This utility metric method is, however, limited in its accuracy, because the amount of energy consumption is so dependent on occupants' preferences and occupation time.

---

[8] The Fanger thermal comfort model will be discussed in Chap. 3.

2.3 Energy as the Main Parameter for Rating Buildings

Energy efficiency is a critical issue for high quality housing. Energy as a measurable variable not only represents a high percentage of the running cost of a building, but also has a major effect on the thermal and optical comfort of the occupants.

In some climates it is difficult to have a comfortable indoor condition without an energy load. As the energy rating of a building can provide specific information on the energy consumption and the relative energy efficiency of the building, it is then possible for a potential buyer to have information on the energy bills that are likely to arise. Through this information the owner of a house may also be able to identify and pinpoint specific cost-effective improvements. However, in a moderate climate a successful passive architectural design could provide thermally comfortable conditions, in which occupants do not need heating and cooling devices. In this case the current energy based rating scheme may fail in its assessment of a building's performance.

Whilst environmental issues were the main reason for developing HERS, financing and marketing have become the major motivations for promoting it. A highly rated building on the market may be eligible for special recognition through a series of voluntary or compulsory programs, which increases its value for sale or rental income. Through HERS, energy-efficient financing is achievable because energy-efficient houses cost less to operate. For the promotion of HERS, the market needs a measurable basis for HERS which is attractive enough for the public to apply for it. Energy and comfort are two parametric options for this purpose, which are related to each other. Energy, as an expensive parameter, would appear to be the more appropriate basis for HERS for marketing purposes, although the provision of comfort may actually be more expensive. However, in modern society in which the public are increasingly dependent on energy for the provision of thermal comfort, energy is seen as a preferable parameter as the basis for HERS.

Connecting HERS and mortgage incentives for energy efficient development has affected the rating systems in the US (National Renewable Energy Laboratory Washington, 1992). HERS provides standardized information on the energy performance of homes, and energy-efficient mortgages (EEM) provide a financing mechanism for energy efficiency. The estimation of energy costs generated by a reliable HERS is a valuable source of information for facilitating EEM. This objective has led to the combining of cost- effectiveness and energy efficiency, and so great attention has been paid to house ratings based on energy usage and its costs.

In addition, predicting ratings on an energy basis helps to choose appropriate HVAC equipment where heating and cooling plants are a part of building construction. This creates an opportunity to optimize heating and cooling plants, and also allows for competition in the market to refine the rated capacity of the size of plant (Hunt, 2003).

The Australian marketing of rating systems is different from that in the US, the EU and Canada. In these countries rating schemes have been employed to support different financial arrangements, while in Australia sustainability and environmental impact are the main policy drivers of the building rating schemes. Moreover in

Australia, with most of its population living in its moderate climate zones, HERS is more amenable to independence from energy and to the provision of thermal comfort as its basis.

2.4 Issues Related to Building Energy Rating Schemes

2.4.1 Rating and Achievement of Sustainability

Current rating schemes have not been sufficiently complex to address the main issues of sustainability. It has been argued in the "design paradigm" that buildings can reverse their environmental impact, and can even have positive impacts over their whole life cycle. This requires integrating conditions for ecosystem preservation in the building fabric. General ecological criteria must then be added to any assessment system for sustainable development. However, current building assessment tools provide only limited support for this issue (Chau et al., 2000). Sustainability is a design problem rather than a technical problem, but the current rating systems are not based on design criteria. Instead, the emphasis is on predicting the negative impacts of a proposed design, such as the level of energy consumption, energy cost and GHG emissions. To move toward sustainable development, Birkenland (2002) proposed that a building must be designed to interact with its context beyond the exterior envelope of the building. It appears that no rating system based on an assessment of energy usage includes all ecologically relevant parameters; not even multi-indicator ratings, such as those described earlier in Sect. 2.2. However, a few include embodied energy, which is a technical aspect that can affect the ecosystem. This is one of the reasons for light-weight buildings being undervalued in the current rating system, while such buildings could actually contribute to improving sustainability.

2.4.2 Rating Free Running Buildings

Free running buildings cannot be accurately evaluated by the current rating schemes. Because all existing rating systems assume buildings to be artificially heated and cooled, they do not deal at all with free running buildings.

When comparing the actual performance of an occupied free running house with the predicted performance by a rating scheme, Soebarto (2000) demonstrated a low score from the rating, although her study shows that the house in question performed reasonably well in terms of its indoor comfort condition, energy use and environmental impact. This reflects the inability of rating systems to assess free running buildings adequately. The benefits of passive architecture design, therefore, may not be properly evaluated, because of the independence of such design from energy use. Another study on the thermal performance of three award winning houses in Australia (Soebarto et al., 2006), illustrated that the houses did not conform to comfort standards and national regulations, in addition to achieving an unacceptable score in the mandated regulatory rating scheme, while at the same time the occupants of all the houses were largely satisfied with the houses' thermal performances.

These two studies imply that there is a difference between an efficient design for a free running house, and that for a conditioned house. This issue is examined in Chap. 4.

2.4.3 Rating Index

Regardless of which method is applied for HERS, an adjusted energy indicator is employed as an indicator of efficient building design. The chosen indicator plays an important role in the reliability of the rating designed to assess the thermal performance of buildings.

Although energy minimization is promoted as an energy efficient building strategy (Boland et al., 2003), low energy usage does not necessarily indicate design efficiency (Sjosten et al., 2003; Olofsson et al., 2004). Energy consumption can be relatively low because the building is not occupied most of the time, or the building amenities are low. Low energy consumption can also be due to efficient appliances. Since appliances consume a significant proportion of the energy used in a home[9] (Environmental Protection Agency, 2000; Office of Energy Efficiency, 2005), highly efficient equipment can reduce the total energy requirement. This means that the energy demand of a building can be reduced by using more efficient appliances, rather than by improving building design.

Furthermore, a *normalized* energy based rating is not sufficient to convey the credibility of an energy efficient design. This point has been argued in many studies (Soebarto, 2000; Williamson, 2000; Meier et al., 2002; Kordjamshidi et al., 2005a). The concept underlying the definition of energy efficient indicators for policy purposes is discussed in Patterson (1996) and Haas (1997). They show that normalized energy use is typically derived as annual energy used, divided by the conditioned floor area or volume. On the basis of this index, a smaller house achieves a poorer value than a similarly constructed larger house (Thomas and Thomas, 2000), where in reality reducing house size is an effective way of reducing total energy consumption (Gray, 1998). One of the reasons for this regressive tendency is a physical phenomenon. Smaller houses have a higher proportion of envelope for a given volume, and therefore the fabric heat flux per unit of floor area or volume is greater in smaller houses. A study of project houses in NSW (SOLARCH, 2000) also found that double storey houses ordinarily could achieve acceptable scores (3.5 stars) with moderate levels of insulation, while single storey houses, especially smaller houses, could not easily achieve this rating. Yet according to one study (Luxmoore et al.,

---

[9] Appliances in a home account for 35% of total energy use on average, and up to 50% in a moderate climate.

2005), the cooling requirements of larger houses with a high energy rating (5 stars or more) were found to be significantly higher than those of houses with a low (3.5) rating, which becomes particularly relevant in the context of predicted global warming (AGO, 2002).

It is most likely that an appropriate indicator for evaluating the efficiency of building design could address the issue of the performance of a building independent of artificial energy load. In that situation, an improvement in the thermal performance of a building should reduce the energy requirements for providing a thermally comfortable space. To fulfil the main objective of HERS, the indicator should be chosen so as to be *related* to the prediction of energy requirements, but not *exclusively based on* a prediction of energy requirements.

If a building is operated in the conditioned mode, the provision of thermal comfort is related to energy consumption. Occupants use energy for space heating or cooling when the indoor climate does not coincide with thermal comfort. However, where the indoor environment is naturally comfortable in terms of temperature and humidity, the need for an active energy load will decrease.

The question then arises as to whether thermal comfort can be used as a basis for assessing the efficiency of a house design. This is important for assessing the efficiency of a house in *entirely* free running operation mode, as opposed to assessing the efficiency of that house in conventional conditioned operation mode on the basis of energy usage.

The correlation between these two bases: comfort and energy, as indicators of the efficiency of a house in different operation modes is addressed in Chap. 4. A probabilistic correlation between thermal comfort and energy requirement does not necessarily mean that a house designed to be free running (comfort based) is an equally efficient conditioned house (energy based). This difference can be crucial with regard to the fundamental role of a house rating system which is intended to influence house performance improvement during the design of a house. This subject will be addressed in more detail in Chap. 4.

2.4.4 Occupancy Scenarios

Almost all reviewed rating systems designed to evaluate the thermal performance of buildings in terms of energy efficiency, set a standard scenario for occupants at the design stage to estimate the annual energy requirements of a building, and then evaluate the thermal performance of the building on the basis of that estimation. However, a standard set of behavioural assumptions for all possible occupancy scenarios cannot give an accurate evaluation.

Occupant behaviour is in fact the most significant determinant of actual energy use. One study suggests that 54% of the variation in energy consumption can be attributed to the building envelope and 46% to occupants' behaviour (Sonderegger, 1978). A similar study (Pettersen, 1994) concluded that where inhabitants' behaviour was unknown, the total predicted energy consumption resulted in +15 to 20% uncertainty, and the range of error for estimated energy heating use was +35 to 40% in a mild winter climate. A number of studies have gone further and shown that actual energy performance depends on the way the occupants "use" the buildings, and does not necessarily relate to the building design at all (Ballinger et al., 1991; Haberl et al., 1998). Indeed, "the predicted energy use or energy cost can be off by 50% or more due to occupant behaviour" (Stein and Meier, 2000).

In a standard occupancy scenario, the parameters such as the number of occupants, period of occupation and thermostat settings for air-conditioners are assumed to be standard. A standard occupancy scenario seems to be essential in order to simplify comparisons of building performance in similar conditions. However different occupancy scenarios can result in different grades or values for a building in a ranking system (Kordjamshidi et al., 2009). Some of the systems provide an option to set the actual number of occupants, but they cannot change the occupied time in a building when they are set for rating the building.

For instance AccuRate software, programmed for HERS in Australia, sets a standard scenario for "occupied time". Living zones are usually considered to be occupied for 17 h a day between 8 a.m. and midnight, and bed zones to be occupied for 17 h between 5 p.m. and 9 a.m. The "17 h scenario" is extremely effective in predicting the thermal performance of a house under a conservative possible occupancy regime, especially when taken together with a completely deterministic estimate of activation of artificial heating and cooling, regardless of occupants' behaviour or climatic seasons.

However, although the occupants' behaviour is not entirely predictable, a more realistic estimation could be employed to evaluate a building's performance and to estimate energy requirements for space heating and cooling. It is not generally possible to predict exactly at which times a dwelling is occupied, but defining multiple occupancy scenarios for rating could result in greater accuracy of prediction.

Setting a single time for occupation can particularly underestimate the value of lightweight buildings. In response to the current concerns about occupancy times and thermostat settings, Boland (2004) noted that "the lightweight dwelling may be disadvantaged unnecessarily". Depending on the time of occupation, a lightweight dwelling may give a better performance because it responds more quickly to temperature changes. This ability, in particular for short period occupation, and particularly in hot summers, is an advantage that cannot be addressed by a permanent "17 h occupancy scenario". The ability of lightweight buildings to achieve a favourable thermal performance needs therefore to be tested for different durations of occupied time.

Occupant behaviours are not a predictable factor. Szokolay (1992a) argues that occupancy factors cannot be taken into account in a rating system because of their high variability; so that the house itself has to be rated. In contrast, Olofsson et al. (2004) argue that if the rating is to reflect the energy efficiency of the occupied building, the actual influence of the users has to be taken into account, for which an evaluation of users is required.

2.4.5 Accuracy of HERS

The credibility of HERS depends on its accuracy. However several studies have demonstrated that the accuracy of energy based rating schemes is questionable. This situation is mainly due to the variability of occupancy behaviour and the rating index, as is described above. While the accuracy of rating systems has not been considered by HERS experts to be the most important barrier to widespread use of HERS, all agree that accuracy is important for the long-term credibility and success of this system. A lack of accuracy may eventually impact on some HERS and cause "irreparable" damage to credibility (Stein, 1997a; Stein and Meier, 2000).

It is clear that while HERS relies on an index of energy, the energy requirement cannot be estimated accurately. A comparison by Stein (1997a) between actual residential energy bills and energy estimation by four different HERS, namely CHEERS, HERO-Ohio, ERHC-Colorado and Midwest-Kansas, demonstrated a significant overestimation (50%) of actual energy cost by CHEERS, and smaller errors in estimating energy cost or energy use by the other methods. However, interestingly, no clear relationship was observed between rating scores and actual energy usage. Stein's case study investigation also showed that it is more difficult to accurately predict energy used in a mild climate than in a severe climate. Stein concluded that the main reason was the variation in occupants' behaviour, and suggested that "incorporating a few pieces of information" about occupants into a rating could improve its accuracy, while elsewhere he pointed out that "actual usage may vary" (Stein, 1997b).

A critical aspect of predicting energy consumption, and consequently of the accuracy of HERS, is determining thermostat settings. All of the current building rating systems consider standard defaults for thermostat settings, taken from thermally comfortable conditions in the building standards, based on a particular strategy. Employing an inappropriate strategy for thermostat settings can effectively reduce the accuracy of predicting energy requirements. This situation has been demonstrated to be more critical in a moderate climate, "where the balance between summer and winter energy consumption is a crucial factor and usually determines the nature of design advice" (Williamson and Riordan, 1997). Neglecting the effect of occupants' behaviour thus also appears to be an issue for thermostat settings in simulation methods for predicting the energy requirements of buildings.

Another issue is that occupants expect a higher degree of comfort in higher scoring buildings, and this tendency results in higher energy consumption than energy usage predicted by rating tools. The discrepancy occurs because the rating system depends on the active energy load, which is variable for different occupants. One way to deal with this problem could be to make house rating schemes independent of energy. Changing the basis of rating from energy to thermal comfort and evaluating buildings in free running mode could encourage the occupants to reduce the energy load for space heating and cooling, and to adapt themselves to natural conditions as far as possible.

2.4.5.1 The Accuracy of HERS Affected by Occupant Seasonal Behaviour

Ignoring seasonal occupant behaviours that respond to the psychological effect of cold and hot months also diminishes the accuracy of HERS. To predict the annual energy requirements in HERS, it is assumed that occupants use energy to maintain indoor temperature in the comfort range whenever the temperature is outside the comfort zone. However, in real life, reasonably, there is no tendency for occupants to mechanically heat a space during summer (hot months) even if the indoor temperature goes down for a few hours. Analogously, the opposite happens for over-heating periods during winter.

In a study by the author (Kordjamshidi et al., 2005b) it was shown that the simulation software correctly predicts that during summer the temperature may come down below the comfort range just between midnight to sunrise, and in winter it may rise above it around midday for just 2 or 3 h. These two particular conditions not only are not critical, but psychologically occupants may accept them as desirable. However, this fact has been ignored in the procedure of calculating or simulating annual energy demand in dwellings in most software developed for HERS, such as NatHERS.

2.5 Need for a New Index for Assessing Building Energy Efficiency

- Rating as ranking
  A rating system requires a simplified method of recognition of the complicated parameters of a building and its occupants. Although estimating energy requirements, particularly through simulation programs, seems a simplified method, this method depends on an active system design for dwellings. Any attempt to achieve an energy efficient design and to reduce energy consumption and GHG emissions relying only on the active energy load to evaluate a dwelling is not going to produce satisfactory results, since it encourages the public to acquire conditioned houses rather than efficient free running ones.

  A reliable rating system would be able to rank buildings in order of the efficiency of their design. This is recognised by Soebarto and Williamson (1999) who claim that "for a HERS mechanism to be sufficient for compliance testing it is only necessary that the scoring system be relatively correct" and by Stein (1997a, p. 17), who argues that "the actual numerical scores are not important as long as the houses are ranked in the correct order". On the other hand, as the above discussions show, it is realized that buildings which are designed for energy conservation in their free running performance cannot achieve a suitable score in the current rating systems. Therefore, when free running and conditioned buildings are ranked in the current rating systems, free running buildings are given inappropriate placement. This occurs when scoring is dependent on energy consumption ratings. There is, therefore, a need for a new index to be introduced, by which the thermal performance of buildings of any design type can be accurately scored and ranked.

To recapitulate, HERS have not been developed to predict the actual energy requirements of a house; the estimation of energy requirements is only a basis on which to make a comparison between the designs of houses for scoring them in relation to energy consumption. Where energy requirements cannot be predicted accurately, the scoring will not be a reliable reflection of the rate of efficiency of houses. If, on the other hand, the efficiency of a house design is to be evaluated on the basis of its free running performance, a new index would need to be proposed as the basis for a House Free running Rating Scheme (HFRS).[10] Where both types of performance of a house, conditioned and free running, are important at the policy level for the development of energy efficiency, then HERS and HFRS should be aggregated within one framework.

- Metrics, norms and diagnostics

  Three elements, namely *metrics*, *norms* and *diagnostics*, are used to evaluate the thermal performance of buildings. Metrics provides a quantification of the performance of the relevant components or systems, without indicating the quality of performance, while they form the basis for developing the norms against which components or system performance are compared. Diagnostics is a procedure involving measurements and analyses to evaluate performance metrics for a system or component under functional testing or actual building site conditions.

  Metrics used for the evaluation of the free running performance of buildings can be derived from the indexes of "thermal comfort". The next chapter reviews thermal comfort criteria to identify how they can be a reliable basis for a house rating scheme.

2.6 Summary

HERS are used to evaluate and promote efficient architectural building design. The most efficient buildings involve architecture design which can provide thermally comfortable indoor conditions for occupants without a mechanical thermal energy load. This means that the efficiency of a building design should be investigated in relation to the thermal performance of the building in free running operation. However, as described in this chapter, energy based ratings cannot at present deal with free running houses. The development of a House Free running Rating Scheme (HFRS), therefore, appears necessary in order to promote efficient architecture design and effectively reduce energy requirements in residential buildings.

---

[10] House Free Running Rating Scheme (HFRS) is a clumsy term in English; however it has been used in this book to make it consistent with the previous term, "House Energy Rating Scheme (HERS)" for house ratings.

With regard to the shortcomings in the current rating schemes (see Sect. 2.4), the following aspects would need to be addressed to develop a reliable and accurate building rating scheme:

- Multiple occupancy scenarios, which should be added to the HERS. This would help to identify the likely better performance of lightweight houses.
- A new index on the basis of thermal comfort should be established as an indicator for evaluating the thermal performance of free running buildings, to form a basis for HFRS.
- The psychological effect of seasons on occupants in computing annual energy requirements should be considered. in order to increase the accuracy of energy based rating systems.
- Comparisons between the thermal performance of houses in conditioned and free running operation mode should be studied to see whether designs for free running houses differ from those for conditioned houses.
- A new framework should be developed for HFRS.
- Since large and double storey houses compared to single storey houses achieve better scores in current HERS, this comparison needs to be tested for free running houses.

These subjects will be addressed in the next chapters in this book, by considering typical residential houses and appropriate tools for evaluating the thermal performance of these houses in different operation modes.

References

AECB.: Minimising CO_2 Emissions from New Homes: a review of how we predict and measure energy use from new homes (2nd edition): Association for Environment Conscious Building, available on line: http://www.aecb.net/ (2006)

AGO.: Understanding Greenhouse Science. http://www.greenhouse.gov.au/ (2002). Accessed 23 Nov 2005

Allen, D.R.: Canada ratings warming up. Home Energy Magazine Online. Available online: http://hem.dis.anl.gov/eehem/99/990910.html (1999)

Andersen, P.D., Jorgensen, B.H., Lading, L., Rasmussen, B.: Sensor foresight–technology and market. Technovation. **24**(4), 311–320 (2004)

Ballinger, J.A.: The 5 star design rating system for thermally efficient, comfortable housing in Australia. Energy Build. **11**(1–3), 65–72 (1988)

Ballinger, J.A.: Towards an Energy Rating Scheme for Residential Buildings in the Northern Territory. Paper presented at a workshop held at Darwin, Australia, 8 May (1991)

Ballinger, J.A.: The Nationwide House Energy Rating Scheme for Australia (BDP Environment Design Guide No.DES 22). Canberra: The Royal Australian Institute of Architects (1998a)

Ballinger, J.A.: The Nationwide House Energy Rating Software (NatHERS) (BDP Environment Design Guide No.DES 23). Canberra: The Royal Australian Institute of Architects (1998b)

Ballinger, J.A., Cassell, D.: Solar efficient housing and NatHERS: an important marketing tool. Proceedings of the Annual Conference of the Australian and New Zealand Solar Energy Society, Sydney, pp. 320–326 (1994)

References

Ballinger, J.A., Samuels, R., Coldicutt, S., Williamson, T.J., D'Cruz, N.: A National Evaluation of Energy Efficient Houses (No.1274 ERDC Project). Sydney: National Solar Architecture Research Unit, University of New South Wales (1991)

Barbara, C.F.: Pilot States Program Report: Home Energy Rating Systems and Energy – Efficient Mortgages (No.NRER/TP- 550- 27722). Colorado: National Renewable Energy Laboratory (2000)

Birkenland, J.: Design for Sustainability: A Sourcebook of Ecological Design Solutions. Earthscan, London (2002)

Boland, J.: Timber Building Construction (N03.1210): The School of Mathematics and Statistics, University of South Australia (2004)

Boland, J., Kravchuk, O., Saman, W., Kilsby, R.: Estimation of thermal sensitivity of a dwelling to variations in architectural parameters. Environ. Modell. Assess. **8**, 101–113 (2003)

Botsaris, P.N., Prebezanos, S.: A methodology for a thermal energy building audit. Build. Environ. **39**(2), 195–199 (2004)

Canada Green Building Council: LEED Canada for home. Canada (2009)

Chau, C.K., Lee, W.L., Yik, F.W.H., Burnett, J.: Towards a successful voluntary building environmental assessment scheme. Constr. Manage. Econ. **18**, 959–968 (2000)

Chung, W., Hui, Y.V., Lam, Y.M.: Benchmarking the energy efficiency of commercial buildings. Appl. Energy. **83**(1), 1–14 (2006)

Cook, G.D., Hackler, R.N., Smith, P.A.: Clothing and laundry techniques to save energy. In Energy Information Handbook (Energy information document 1028): University of Florida (1997)

Department of Environment, Food and Rural Affairs (DEFRA): The Government's Standard Assessment Procedure for Energy Rating of Dwellings. BRE, Garston, Watford. www.bre.co.uk/sap2005 (2005). Accessed 15 Jan 2006

Energies-Cites: Energy Management in Municipal Buildings. http://www.display-campaign.org/IMG/pdf/case-study_odense_en.pdf (2003). Accessed 10 Nov 2007

Energy Efficiency Partnership for Homes: Measuring up the Home Energy Ratings. http://www.est.org.uk/partnership/energy/lead/index.cfm?mode=view&news_id=559 (2006). Accessed 10th Aug. 2008

Environmental Protection Agency: Energy Efficient Appliances (No. EAP 430-F-97-028). US: EAP (2000)

Fairey, P., Tait, J., Goldstein, D., Tracey, D., Holtz, M., Judkoff, R.: The HERS Ratings Method and the Derivation of the Normalized Modified Loads Method (No. FSEC-RR54-00). Florida: Florida Solar Energy Centre, Cocoa (2000)

Gellender, M.: Energy Rating and/or Energy efficiency standards for new houses: issues and options for Queensland: Presented at a workshop sponsored by the Queensland Energy Information Centre (1992)

Gray, E.: NatHERS Effect of Dimension on Star Energy Rating (A report prepared for WA Office of Energy) (1998)

Haas, R.: Energy efficiency indicators in the residential sector: what do we know and what has to be ensured? Energy Policy. **25**(7–9), 789–802 (1997)

Haberl, J., Bou-saada, T., Reddy, A., Soebarto, V.: An evaluation of residential energy conservation option using side-by-side measurements of two habitats for humanity houses in Houston, Texas, Proceedings of the 1998 ACEEE Conference, American Council for an Energy Efficient Economy, California (1998)

Hasson, F., Keeney, S., McKenna, H.: Research guidelines for the delphi survey technique. J. Adv. Nurs. **32**(4), 1008–1015 (2000)

Hunt, S.: Focus on Construction Quality, Monthly Newsletter US: IBACOS (2003)

International Energy Agency: IEA ECBCS Annex 36: Energy Concept Adviser for Technical Retrofit Measures- Energy Audit Procedures. (Jan de Boer ed) December (2003)

Kordjamshidi, M., Khodakarami, J., Nasrollahi, N.: Occupancy scenarios and the evaluation of thermal performances of buildings, Proceeding of ANZSES conference, Townsville, Australia (2009)

Kordjamshidi, M., King, S., Prasad, D.: An Alternative Basis for a Home Energy Rating Scheme (HERS). Proceedings of PLEA, Environmental sustainability: the challenge of awareness in developing societies, Lebanon, pp. 909–914 (2005a)

Kordjamshidi, M., King, S., Prasad, D.: Towards the Development of a Home Rating Scheme for Free Running Buildings. Proceedings of ANZSES, Renewable Energy for a Sustainable Future- A challenge for a post carbon world. New Zealand. Dunedin University (2005b)

Kotsaki, K. and Sourys, G.: Critical Review and State of the Art of the Existing Rating and Classification Techniques. Group Building Environmental Studies, University of Athens, Athens (2000)

Luxmoore, D.A., Jayasinghe, M.T.R., Mahendran, M.: Mitigating temperature increases in high lot density sub-tropical residential developments. Energy Build. **37**(12), 1212–1224 (2005)

Meier, A., Olofsson, T., Lamberts, R.: What is an energy-efficient building? Proceedings of the ENTAC 2002- IX Meeting of Technology in the Built Environment, Brazil (2002)

Miguez, J.L., Porteiro, J., Lopez-Gonzalez, L.M., Vicuna, J.E., Murillo, S., Moran, J.C. et al.: Review of the energy rating of dwellings in the European union as a mechanism for sustainable energy. Renewable Sustain Energy Rev. **10**(1), 24–45 (2006)

Mills, E.: Inter-comparison of north American residential energy analysis tools. Energy Build. **36**(9), 865–880 (2004)

National Renewable Energy Laboratory Washington: A National Program for Energy-Efficient Mortgages and Home Energy Rating Systems: A Blueprint for Action (No. NREL/TP-261-4677), Washington, DC (1992)

Office of Energy Efficiency: The State of Energy Efficiency in Canada (No. M141-7/2004). Canada: The Office of Energy Efficiency of Natural Resources (2005)

Olofsson, T., Meier, A., Lamberts, R.: Rating the energy performance of buildings. Int. J. Low Energy Sustain. Build. **3**, 1–18 (2004)

Patterson, M.G.: What is energy efficiency? Concepts, indicators and methodological issues. Energy Policy. **24**(5), 377–390 (1996)

Pettersen, T.D.: Variation of energy consumption in dwellings due to climate, building and inhabitants. Energy Build. **21**(3), 209–218 (1994)

Richalet, V., Henderson, G.: Europe Union Not Unified on Home Ratings. Home Energy Magazine Online. http://hem.dis.anl.gov/eehem/99/990911.html (1999, Sep/Oct). Accessed 17 Sep 2007

Richalet, V., Neirac, F.P., Tellez, F., Marco, J., Bloem, J.J.: HELP (house energy labelling procedure): methodology and present results. Energy Build. **33**(3), 229–233 (2001)

Roulet, C.-A., Flourenttzos, F., Santamouris, M., Koronaki, I., Daskalaki, E., Richalate, V.: ORME-Office Building Rating Methodology for Europe (Office Project Report). University of Athens (1999)

Roulet, C.-A., Flourentzou, F., Labben, H.H., Santamouris, M., Koronaki, I., Dascalaki, E. et al.: ORME: A multicriteria rating methodology for buildings. Build. Environ. **37**(6), 579–586 (2002)

Roulet, C.A., Johner, N., Oostra, B., Foradini, F., Aizlewood, C., Cox, C.: Multi-criteria analysis of health, comfort and energy efficiency of buildings, The 10th International Conference on Indoor Air Quality and Climate, Beijing, pp. 1174–1178 (2005)

Santamouris, M.: Energy Retrofit of Office Buildings. Athens: CIENE; University of Athens (1995)

Santamouris, M.: Energy Performance of Residential Buildings: A Practical Guide for Energy Rating and Efficiency. James & James, Earthscan, UK, USA (2005)

Santamouris, M., Dascalaki, E.: Passive retrofitting of office buildings to improve their energy performance and indoor environment: the OFFICE project. Build. Environ. **37**(6), 575–578 (2002)

Santamouris, M., Mihalakakou, G., Patargias, P., Gaitani, N., Sfakianaki, K., Papaglastra, M. et al.: Using intelligent clustering techniques to classify the energy performance of school buildings. Energy Build. **39**(1), 45–51 (2007)

References

Sjosten, J., Olofsson, T., Golriz, M.: Heating energy use simulation for residential buildings, Eight International IBPSA Conference, Eindhoven, Netherlands, pp. 1221–1226 (2003)

Soebarto, V.I.: A Low-Energy House and a Low Rating: What is the Problem, Proceedings of the 34th Conference of the Australia and New Zealand Architectural Science Association, Adelaide, South Australia, pp. 111–118 (2000)

Soebarto, V.I., Williamson, T.J.: Design orientated performance evaluation of buildings, Building Simulation '99. Sixth International IBPSA Conference, Kyoto, Japan. International Building Performance Simulation Association, pp. 225–232 (1999)

Soebarto, V.I., Williamson, T.J.: Multi-criteria assessment of building performance: theory and implementation. Build. Environ. **36**(6), 681–690 (2001)

Soebarto, V., Williamson, T., Radford, A., Bennetts, H.: The performance of award winning houses, The 23rd Conference on PLEA, Geneva, Switzerland, pp. 855–860 (2006)

SOLARCH: Project Homes: House Energy Rating, New South Wales Industry Impact Study (A report prepared for the Sustainable Energy Development Authority): University New South Wales (2000)

Sonderegger, R.C.: Movers and stayers: the resident's contribution to variation across houses in energy consumption for space heating. Energy Build. **1**(3), 313–324 (1978)

SRC: Review of Home Energy Rating Schemes: Findings and Recommendation (No. 03-412-8900). Melbourne, Victoria: SRC Australia Pty Ltd (1991)

Stein, J.R.: Accuracy of Home Energy Rating Systems (No. 40394). US: Lawrence Berkeley National Laboratory (1997a)

Stein, J.R.: Home Energy Rating Systems: Actual Usage May Vary. Home Energy Magazine Online. http://hem.dis.anl.gov/eehem/97/970910.html (1997b, Sep/Oct). Accessed 10 May 2007

Stein, J.R., Meier, A.: Accuracy of home energy rating systems. Energy. **25**(4), 339–354 (2000)

Szokolay, S.: An energy rating system for houses. In Energy-efficient Ratings and Standards for New Houses. Brisbane: Queensland Energy Information Centre Department of Resource Industries (1992a)

Szokolay, S.V.: HERS: Proposal for a Nationwide Home Energy Rating Scheme (report to Dept. of Primary Industries and Energy) (1992b)

Thomas, P.C., Thomas, L.: A study of an energy consumption index normalised for area in house energy rating schemes. Proceedings of the 38th Annual Conference of the Australian and New Zealand Solar Energy Society: From Fossils to Photons Renewable Energy Transforming Business, Brisbane, pp. 113–121 (2000)

Turrent, D., Mainwaring, J.: Saving energy on the rates. RIBA J. 85–86 (1990, September)

US Department of Energy: Model Energy Code Compliance Guide Version 2.0: Us Department of Energy Building Standards and Guidelines Program (1995)

US Department of Energy: Building Energy Software Tools Directory. http://apps1.eere.energy.gov/buildings/tools_directory/subjects_sub.cfm (2009). Accessed 13 Nov 2009

Wathen, G.: Energy-efficient Rating Schemes and Building Standards in Victoria. In Energy-efficient Ratings and Standards for New Houses. Papers presented at a workshop sponsored by the Queensland Energy Information Centre, April 29, 1992, pp 1–16 (1992)

Williamson, T.J.: A critical review of home energy rating in Australia, Proceedings of the 34th Conference of the Australia and New Zealand Architectural Science Association, Adelaide, South Australia, pp. 101–109 (2000)

Williamson, T., Riordan, P.: Thermostat strategies for discretionary heating and cooling of dwellings in temperate climates. Proceeding of 5th IBPSA Building simulation Conference, Prague: International Building Performance Simulation Association, pp. 1–8 (1997)

Zmeureanu, R., Fazio, P., DePani, S., Calla, R.: Development of an energy rating system for existing houses. Energy Build. **29**(2), 107–119 (1999)

Chapter 3
Thermal Comfort

The question of how to establish thermal comfort as a basis for HRS is a broad subject. The extent to which a dwelling can provide thermally comfortable conditions for its occupants is determined from the difference between prevailing weather conditions and the desired comfort condition. The desired comfort condition is therefore determined in the context of climate.

This chapter reviews general aspects of thermal comfort and approaches to measuring thermal comfort for the assessment of naturally ventilated buildings. It proposes an appropriate indicator as a basis for building performance assessment in the free running operation mode, and then establishes thermal comfort boundaries in specified moderate climates to be used in the evaluation of the free running performance of houses.

3.1 Thermal Comfort

The main objective of any effort to develop energy efficient buildings is the provision of thermal comfort with minimum energy consumption, by employing climatic building design. One of the key points is to ensure that occupants' thermal comfort is not sacrificed in order to reduce energy requirements. Therefore any assessment system to evaluate the efficiency of a particular building needs to consider the criterion of thermal comfort as the context for its evaluation of building performance.

The objective of this section is to substantiate the criterion of thermal comfort for evaluating the thermal performance of buildings. It first reviews many definitions of thermal comfort in the building sector to determine the most appropriate one for this study. Secondly, it reviews the main variables which impact on the provision and affect the sensation of thermal comfort, in order to specify the main parameters that should be considered in evaluating the performance of a building on this basis. The third and more important section is a review of the current standards, in order to select an appropriate indicator of thermal comfort on which building performance can be evaluated.

3.1.1 Definition of Thermal Comfort

Numerous definitions of thermal comfort have been proposed by various researchers (Fanger, 1970; Givoni, 1976; Watt, 1963; O'Callaghan, 1978; Benzinger, 1979; Hensen, 1990; Ihab, 2002; Chappells and Shove, 2004). Generally thermal comfort is defined as "that condition of mind which expresses satisfaction with the thermal environment" (ASHRAE, 2001). Owing to biological variance, it is not possible for everyone to be satisfied at the same time and in the same climate. Therefore some subjective criteria need to be applied in order to establish optimal comfort. Fanger (1970) suggests an optimal thermal condition as being one in which the highest possible percentage of a group is in thermal comfort. Optimal thermal condition is defined as a state in which there is no driving impulse to correct the environment by behaviour. Thermal comfort is interpreted by Givoni (1976) as the absence of irritation and discomfort due to heat or cold, and the state of pleasantness. Hensen (1990) detailed the causes of dissatisfaction in terms of the whole body's being too warm or cold, as well as unwanted heating or cooling in a part of the body (local body). Benzinger (1979) examined thermal comfort in terms of "ideal" thermal comfort and relative or "mixed" comfort.

A universal definition of comfort is almost impossible because of people's variable preferences, and the particular characteristics of different climates which affect the sensation of thermal comfort. A condition in which 80–90% of occupants feel thermally comfortable is accepted as a standard universal term. The specification of that condition depends on the parameters or variables which impact on thermal comfort. Theses parameters are described in the following section.

3.1.2 Human Comfort and Variables Affecting Thermal Comfort

Human life requires a deep – body temperature of between 35 and 40°C, for which 37°C is the proxy mean. Skin temperature is normally between 31 and 34°C. If heat is to be lost, then the temperature of the surrounding environment must be less than skin temperature.

The most important variable to determine human comfort is air temperature. However this is not the only indicator. A number of factors influence the various heat exchange processes on the body surface which affect the sensation of comfort or discomfort (Szokolay, 1980). These factors are divided into two groups: environmental "climatic" and "non-climate" factors. The main environmental parameters have been identified as:

- Air temperature
- Humidity
- Air velocity
- Radiation (Parsons, 2003; Szokolay, 2004)

3.1 Thermal Comfort

In addition, there are other factors involved, such as draught, a high vertical temperature difference between head and ankles, or too high radiant temperature asymmetry. Increasing temperature always causes a corresponding change in the thermal sensation.

The impact of humidity on the human thermal comfort balance is complex. It has no significant effect on thermal comfort unless temperatures are very high or low. It has been demonstrated that up to about 27°C a sedentary subject could not experience any difference between relative humidities of 30 and 80% in their subjective sensations (Givoni, 1998). At a comfortable temperature perspiration is not important, but the heat dissipation mechanism is important at high temperatures.

The rate of evaporation of perspiration depends on the absolute humidity of the surroundings. High proportions of humidity (above 12 g/kg) can cause unpleasant sensations because of restricting evaporation and consequently its cooling effect. Low humidity (generally less than 4 g/kg) can cause drying out of mucous membranes. Thus the effect of humidity on the sensation of thermal comfort cannot be ignored in a climate, in which the relative humidity (RH) is higher much of the time than the acceptable range.

Air movement around the body convectively transfers heat and causes cooling. Still air surrounding the body produces a thin insulation layer around the body. Air movement reduces the thickness of this insulation and so provides a cooling effect. It is an important mechanism for removing heat generated by the body, particularly when the level of humidity is high. Increasing air velocity decreases the amount of moisture held in the air around the body and results in increasing evaporation (ASHRAE, 1981).

The beneficial effect of air velocity therefore should not be ignored in an efficient design, particularly in a humid climate, as it can significantly reduce cooling energy requirements. This is a key point in designing for free running buildings. The employment of a specific strategy to provide maximum air circulation in indoor spaces should be one of the design priorities for free running buildings in a moderate climate with relative high humidity.

The thermal sensation experienced by a subject in an environment is significantly affected by the radiative heat exchange between the human body and the surrounding surfaces. This contributes as much as 30% of the whole thermal exchanges of the subject (La Gennusa et al., 2005).

Mean Radiant Temperature (MRT) is used to define the average temperature of all surfaces in a given space to which the body is exposed. The MRT is twice as important as the dry bulb temperature for lightly clothed people, while in cooler climates these two temperatures are equally important. Olgyay (1963) in his bioclimatic chart showed the interaction of the four main environmental factors for thermal comfort, and used a 0.8°C increase in the mean radiant temperature to adjust to a 1°C decrease in the dry bulb temperature. This chart, however, has been revised and has become more complicated in later more sophisticated studies. The most well-known is found in the ASHRAE standard.

The interaction between the above four climatic variables is what makes an indoor environment thermally comfortable or uncomfortable. Givoni (1998) notes that heat discomfort inside buildings is correlated generally with "environmental temperature" and air speed over the body. Environmental temperature is the combined effect of the air temperature and mean radiant temperature of the enclosure. If the air and mean radiant temperature are not the same, the globe temperature is a convincing measure of the resulting environmental temperature. Environmental temperature would therefore appear to be a more appropriate indicator than only air temperature as one of the factors for evaluating the degree of comfort of indoor conditions.

There are certain other, "non-climate", factors which can affect how comfortable a person feels in a given situation, such as: age (Mayer, 1993; Young, 1991; Young and Lee, 1997), clothing, acclimatisation, sex (Chung and Tong, 1990; Lee and Choi, 2004; Nakano et al., 2002; Parsons, 2001), health and activity (Parsons, 2003; Yoshida et al., 2000), and subcutaneous fat.

While the geographic location would seem to have an influence on thermal sensation, Parsons (2003) argues that this has not been shown to be the case in some research. He refers to the observation of various studies (Ellis, 1953; Fanger, 1970) that found no significant difference between conditions preferred by subjects in different geographical locations. However, it should be noted that the results of those studies were obtained for occupants in conditioned buildings and not free running, naturally ventilated ones, and, as will be explained later, in the case of the latter geographical location is relevant.

The sensation of comfort depends on the activity, physiology and thermoregulatory system of the body (Gagge et al., 1937; Yaglou, 1927; Winslow, Herrington and Gagge, 1937; Fanger, 1967; Gagge et al., 1967; McNall et al., 1967; Gagge, 1973). The level of an occupant's activity can be roughly predicted by noting a building's function. For instance, in office buildings occupants can be considered to be sedentary with a low level of activity (1–1.2 met). However, the variation of activity in residential buildings is not predictable. Because of this unpredictability and the type clothing worn in the residential sector, the criteria for comfort conditions for this type of building may differ from those in buildings such as offices. It was observed by de Dear et al. (1997) that although there were distinct differences in the degree of behavioural thermoregulatory adjustment made by residential building occupants compared to those in office buildings, there were no discernibly sharp differences in occupants' evaluations of the building's indoor climatic quality. Thus, in at least some situations, the criteria of thermal comfort for residential buildings can be considered to be similar to those in office buildings.

3.1.3 Thermal Comfort Models and Standards

Over the past 100 years many research efforts have been devoted to developing indexes and models predicting the thermal sensation of people. Thermal comfort prediction models generally are mathematical models of the relationship between one or more environmental factors and certain occupant factors. The main aim of comfort models is to provide a single index that encompasses all relevant parameters.

Thermal models of the human body and its interaction with the surrounding thermal environment are often proposed and used as the basis of thermal comfort standards. Comfort standards rely on such models of human thermal comfort to establish the interior environmental conditions they prescribe, and to provide a single index that encompasses all the relevant physical parameters. The two basic types are empirical and theoretical. A review of the development of these two types and their details for measuring thermal comfort has appeared in "Thermal Comfort" (Auliciems and Szokolay, 1997) and Proceedings of "Moving Thermal Comfort Standards into the twenty-first century" (2001). The models that have been developed vary from a simple linear equation relative to an indoor comfort temperature, to outdoor dry bulb temperature, to complex algorithms. Both simple and complex models have limitations for use in establishing standards. These limitations affect the accuracy of any system, such as a simulation that employs such models, as well as affecting the accuracy of the inputs to the model (Jones, 2001). The point is that the most complex models are not always the most accurate, and the simplest are not always the easiest to use. However, the simpler index is most likely to find widespread practical application (Holm and Engelbrecht, 2005). The accuracy of the model depends to what purpose it has been used for.

The most notable models have been those developed by Fanger (1967, 1970), the Pierce Foundation (the Pierce two-node mode) originally developed by Gagge et al. (1971), and that of researchers at Kansas State University (The KSU two-node model) (Azer and Hsu, 1977). The theory behind these three models is described by Berglund (1978). All three models are based on the concept of an energy balance in a person, and use the mechanism of energy exchange, along with physiological parameters that were derived experimentally to predict the thermal sensation and the physiological response of a person to their environment. The models differ in the criteria which are used to predict thermal sensation, physiological models for heat transfer from the body, and the human control system. Among these three models, Fanger's model appears to be the most commonly used in academic research, the development of software and the establishment of standards.

ASHRAE standard 55 (2004) and ISO 7730 (1995) have been widely adopted as international thermal comfort standards. These standards are based on a human energy balance obtained by assuming steady-state conditions. Those are deduced from the experiments conducted by Fanger (1970) in climatic chambers, using predicted mean vote (PMV) and Predicted Percentage Dissatisfied (PPD) to estimate the human mean response to the thermal environment from six thermal variables. These related indices are based on a combination and interaction between environmental and personal parameters as follows:

- Environmental parameters comprise:
  - air temperature
  - radiant temperature or globe temperature

- air velocity
- air relative humidity (RH) or vapour pressure

• Personal parameters that are related to occupant adaptability to the local climate comprise:

- metabolic rate
- clothing insulation

However, it has been demonstrated that ISO 7730 and PMV/PPD overestimate warm discomfort (de Dear et al., 1997; Humphreys and Nicol, 1995; Karynono, 1996; Williamson et al., 1995).

ASHRAE provides the recognized world standard for thermal comfort in interior environments. This standard sets a narrow temperature zone in which 80–90% of slightly active people would find the environment thermally acceptable. However thermal discomfort is often reported by a large percentage of occupants in offices when the thermal environment complies with the recommendations in the standards (Melikov, 2004). This is related to the variability of occupants' preferences. Personality differences in preferred air temperature may be as great as 10°C (Grivel and Candas, 1991). Occupants' preferences for air movement may differ by more than four times (Melikov, 1996).

A thermal comfort zone is typically determined on a psychometric chart which is related to the air temperature and humidity. A combination of humidity and air temperature to determine the breadth of the comfort zone appears in (ET*) index (ASHRAE, 2003).

3.1.4 Applicability of the Thermal Comfort Index for Naturally Ventilated Buildings

Arguably the most widely accepted index of thermal sensation is Fanger's "predicted mean vote" (PMV) (1970), which is the main index of comfort in ASHRAE Standard 55. This comfort index was developed on the basis of the physics of heat transfer, combined with an empirical fit to sensation, and based on the steady state. It is known to be a complicated equation because of the need to consider the main factors (personal and environmental) affecting thermal comfort. Although it is the most appropriate thermal index for buildings with an environmental control system, there are some features which limit its applicability. The PMV model does not include the effect of solar radiation through windows on the occupant. It only includes the mean radiant temperature of a space in its computation. Thus discomfort caused by window radiation cannot be predicted by this model. This problem is addressed by Lyons et al. (1999; 2006), who proposed a solar correction factor when calculating PPD.

The steady state condition, on the basis of which the PMV model was developed, makes it inapplicable to free running houses. This has been demonstrated by many studies, particularly by de Dear (Baker and Standeven, 1996; Brager and de Dear, 1998; de Dear et al., 1997; Forwood, 1995; Humphreys, 1975; Nicol and Aulicien, 1994; de Dear and Brager, 1998; Brager and de Dear, 2000; Brager and de Dear, 2001; de Dear and Brager, 2001; de Dear and Brager, 2002; Humphreys and Fergus Nicol, 2002; de Dear, 2004). These studies share many of the concerns about the inapplicability of a laboratory based index to free running buildings.

Because of the strict application of steady state conditions, the index exaggerates the percentage of dissatisfied people if it is used for naturally ventilated buildings, and where indoor temperature is controlled manually by the occupants according to their feelings. People in real situations show a wider range of preferences than they do in laboratory experiments. A review of thermal comfort standards by Lovins (1992) showed that the comfort model developed from chamber research is "seriously flawed", basically because it overlooks factors such as acclimatization, dependence and physiological variables among individuals. It has also been argued that strict reliance on laboratory-based comfort standards ignores important contextual influences that can decrease sensitivity to a given set of thermal conditions (Brager and de Dear, 1998). A number of case studies (Bouden and Ghrab, 2005; Feriadi and Wong, 2004; Wong et al., 2002) have also shown a significant deviation between thermal comfort sensation in naturally ventilated buildings and what PMV predicts. From an exhaustive analysis of all reported research from both naturally ventilated and HVAC controlled buildings, de Dear et al. (1997) concluded that while a mechanistic model of heat transfer may well describe the responses of people within a controlled indoor thermal environment, it is "inapplicable to naturally ventilated premises because it only partially accounts for processes of thermal adaptation to indoor climate".

The heat balance model ignores the psychological adaptation of occupants to a natural climate, and the fact that the tolerance of occupants in free running building is wider than in conditioned buildings. Thermal sensation, satisfaction, and acceptability are all influenced by the match between one's expectations about the indoor climate in a special context and what the actual outdoor environment is (Fountain et al., 1996). de Dear and Hart (2002, p. 1) sums this up by stating that this model "...ignores the psychological dimension of adaptation which may be particularly important in contexts where people's interactions with the environmental (i.e. personal thermal comfort), or diverse thermal experiences, may alter their expectations, and thus their thermal sensation and satisfaction. One context where these factors play a particularly important role is in naturally ventilated buildings."

de Dear and Hart (2002) points out that the environmental inputs to conventional heat balance thermal comfort models, such as PMV, have been taken from the indoor environment surrounding the building occupants. Such models also need the user to have information on the occupants' clothing insulation (clo) and metabolic rates. These last parameters are often difficult to estimate in the field, particularly for the use of free running house rating, in which we need to consider identical conditions for occupants in order to obtain a reliable comparison between dwellings.

A study by Brager and de Dear (2000) clearly showed that thermal sensation based on PMV does not correspond with that for naturally ventilated buildings. This study developed the following model, which shows how people felt too warm or too cool in conditioned buildings as compared to naturally ventilated buildings. It was found that the occupants of centralized HVAC buildings were twice as sensitive to deviations in temperature as were occupants of naturally ventilated buildings.

$$TS - 0.51\ T_{op} - 11.96\ \text{(Centralized HAVC buildings)} \tag{3.1}$$

$$TS = 0.27\ T_{op} - 6.65\ \text{(Naturally ventilated buildings)} \tag{3.2}$$

In which:

TS = mean thermal sensation, which represents a vote on the seven point thermal sensation (PMV).
T_{op} = mean indoor operative temperature

Figure 3.1 illustrates a comparison between the thermal sensation of occupants of a naturally ventilated building and a building with centralized HAVC, based on the Brager models. It reveals a greater difference between the thermal sensations in the discomfort temperature range. It would appear from her comparison that any model developed for evaluation of the thermal performance of a conditioned building cannot be applied for the evaluation of a free running building.

An extension of the PMV model that includes an "expectancy" factor was added by Fanger and Toftum (2002) to the PMV index, to make it applicable for use in non-air-conditioned buildings in warm climates. This model accorded well with some field studies in warm climates, but its applicability for other climates needs to be examined further. To extend such a PMV model for naturally ventilated buildings, more research needs to be done before any practical implications can be drawn.

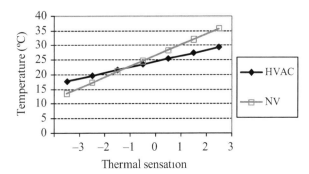

Fig. 3.1 Comparison of PMV for conditioned and free running buildings based on the de Dear study

3.1.5 Adaptive Thermal Comfort Models for Naturally Ventilated/Free Running Buildings

The adaptive thermal comfort index has been developed through several investigations using "real" people engaged in "real" tasks in "real" built environments rather than in laboratory experiments. A number of studies have shown a correlation between outdoor temperature and thermally comfortable indoor condition for naturally ventilated buildings. "Meta-studies" of thermal comfort field studies have shown that indoor comfort temperature as felt by the occupants is a function of the mean outdoor temperature (Auliciems and de Dear, 1986; Brager and de Dear, 1998; de Dear et al., 1993; Nicol and Aulicien, 1994; Nicol and Roaf, 1996; de Dear and Brager, 1998). This means that we can relate indoor comfort temperature to climate, region and seasons. For free running buildings and according to different surveys conducted under different climatic conditions, Humphreys (1976), reviewing the available field data, found a strong statistical dependence of thermal neutrality on the mean level of air or globe temperature. He found (1978) that the comfort temperature can be obtained from the mean outdoor temperature with Eq. (3.3)

$$T_n = 0.534T + 11.9 \tag{3.3}$$

Auliciems (1981) revised Humphreys' equation by deleting some field studies, such those with children as the subjects, and adding more information from other studies not included by Humphreys. These revisions increased the database to 53 separate field studies in various climatic zones, covering more countries and more climates. After combining the data for naturally ventilated and air-conditioned buildings, the analysis led to an equation involving both the outdoor air temperature (T_o) and the indoor air temperature (T_i). The resulting equation is (Eq. 3.4):

$$T_c = 0.48T_i + 0.14T_o + 9.22 \tag{3.4}$$

Auliciems and de Dear (1986) have also proposed a single line for all buildings, covering naturally ventilated buildings and air-conditioned buildings. This relation is given by Eq. (3.5)

$$T_c = 0.31T_o + 17.6 \tag{3.5}$$

Nicol and Roaf (1996) has conducted several surveys under different climatic conditions. In a first survey in Pakistan he established a relation between comfort temperature and outdoor temperature given by Eq. (3.6).

$$T_c = 0.38T_o + 17.0 \tag{3.6}$$

In a second survey in Pakistan (Nicol et al., 1999), he developed a second regression given by Eq. (3.7).

$$T_c = 0.36 T_o + 18.5 \qquad (3.7)$$

These relations show clearly that the comfort temperature is related to the outdoor temperature and so to the climate. A regression has been developed in the function of outdoor ET*(Effective Temperature) (de Dear et al., 1997). The equation for all buildings is:

$$T_n = 20.9 + 0.16 \, ET^* \qquad (3.8)$$

And for free running buildings is:

$$T_n = 18.9 + 0.255 \, ET^* \qquad (3.9)$$

According to De Dear and Hart (2002) the adaptive comfort model which was formulated in terms of mean monthly outdoor air temperature is more applicable and familiar than ET* to engineers.

> ....It was agreed by every one on SSPC 55[1] that ET* is primarily an index used by researchers, and that practitioners would be more likely to use ACS[2] if the meteorological input data was a more familiar and accessible index. The ACS was therefore reformulated in terms of mean monthly outdoor air temperature, defined simply as the arithmetic average of the mean daily minimum and main daily maximum outdoor (dry bulb) temperatures for the month in question. This climate data is readily available and familiar to engineers. (De Dear and Hart, 2002, p. 557).

$$T_{comf} = 0.31 T_{a,out} + 17.8 \qquad (3.10)$$

The above studies have demonstrated a line of best fit through data analyses. Figure 3.2 collects these lines together. Variations can be seen between these studies, particularly between the first study by Humphreys and the last work, done by de Dear. The two lines intersect at a thermal neutrality of 25°C. These measures of temperature are only the same when the relative humidity is 50%, but a discrepancy exists at other levels of humidity. The reasons for this lie in the type of buildings and the characteristics of occupants, such as their physiological, psychological and cultural features. Although the field investigations seem to cover different countries, this particular issue needs more research to show the probability of the effect of culture on thermal sensation and energy consumption.

Although the best correlation is shown between the adaptive comfort models and thermal sensation in naturally ventilated buildings, a more complex index in which all effective environmental and personal factors would be included, needs to

---

[1] SSPC 55 is the ASHRAE committee in charge of revising thermal comfort standards.
[2] ACT Adoptive Comfort Standard.

3.1 Thermal Comfort 41

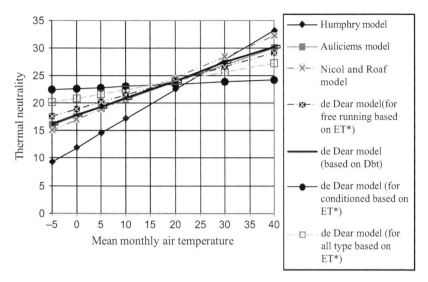

Fig. 3.2 Thermal neutrality models, which show the correlation between thermal neutrality and mean monthly outdoor temperature (DBT)

be developed for free running houses. This model does not include human clothing or activity, nor the four classical thermal parameters that have a significant impact on the human heat balance and therefore on thermal sensation. However, the model should be applicable where there is no other completed model developed for free running buildings.

3.1.6 Acceptable Thermal Conditions in Free Running Buildings Based on the ASHRAE Standard

ASHRAE (2004) introduced an acceptable operative temperature range for naturally conditioned buildings, based on de Dear's adaptive model. It is applicable for spaces with operable windows that can be opened to the outdoors and adjusted by the occupants. In this model, metabolic rates range from 1.0 to 1.3 met. Although "no humidity or air limits are required" in the application of this model, one cannot ignore the effect of humidity and air ventilation on the sensation of thermal comfort, particularly in warm and high humid climates.

3.1.7 Applicability of the Adaptive Comfort Model for Free Running Residential Buildings

The applicability of the adaptive comfort standard in residential buildings is a challenge, since defining the criteria of thermal comfort in such buildings is problematic

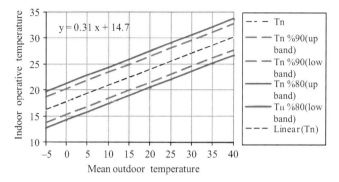

Fig. 3.3 Adaptive thermal comfort standard (ACS), applied for naturally ventilated buildings in ASHRAE 55-2004. Based on (ASHRAE, 2004)

owing to the substantial variability of occupants' behaviour. The adaptive comfort standard has emerged from many studies done basically on office buildings, in which the scenario of occupancy in terms of occupation time, of where occupants sit, and of the level of activity and of clothes is predictable. However, due no doubt to the wide variety of occupant behaviour in residential buildings, no comfort model has been developed specifically for residential buildings.

A study by the Davis Energy Group (2004) presents the reasons for which the thermal performance of a residential building are most likely to differ from the developed standards, including adaptive models. The differences are said to be owing to the following factors:

- activity
- size of population
- steady-state assumption
- assumption of natural ventilation or HVAC but not a combination of both
- minimisation of "circumstantial restraints".

The report highlights that "finding some agreement on input conditions to generate a comfort zone for more than one conditioned zone and for all occupants is a challenge" (Davis Energy Group, 2004, p. 7). While a comfort model is developed for the sedentary activity of office work, it does not apply to bed zones or to children, the elderly and disabled people, who would be considered as among the occupants. Furthermore, the number of occupants varies for different families, which may influence the thermal conditions of how a house performs. Individuals for many different reasons vary in their comfort preferences and all variations cannot be predicted for residential buildings. A thermal condition of a house "tends to cycle through great flux in internal gains and external gains". However, a standard cannot address the effect of this cycle on thermal comfort in adaptive models. An adaptive model is only suitable for buildings with no mechanical condition at all; it is not applicable for houses with combined systems. Occupants in residential buildings have a wide flexibility in choice for clothing, activity and location to adjust themselves to indoor conditions in order to become comfortable. The concept "circumstantial restriction", which is described by Humphreys and Nicol (1998), and considered by them to develop comfort models for predominantly non-domestic settings, may not be observed in houses.

In spite of the above argument, the adaptive comfort model appears to be more applicable for free running houses than other thermal comfort models, since there is no established model developed for residential buildings. The main advantage of the adaptive model which makes it applicable for free running houses is that it respects the effect of acclimatization. Acclimatization is the main parameter that influences the evaluation of thermal performance of a house through its effect on the behaviour of occupants. This might be a reason for differences between the evaluations of thermal performance of a house in different operation modes

3.2 Evaluation of a Residential Building's Thermal Performance on the Basis of Thermal Comfort

The evaluation of the thermal indoor climate of a building in terms of human comfort response can be classified by the percentage of satisfied or dissatisfied occupants. This method has been employed in ISSO (2004) and two other studies (Olesen et al., 2006; van der Linden et al., 2006) that categorise buildings into three different groups: A, B and C. Level A corresponds to 90% thermal acceptability, and is applied to buildings with high performance for thermal comfort. Level B is defined as corresponding to 80% thermal acceptability, meaning good indoor thermal comfort, and finally 65% thermal acceptability is labelled level C and can be applied in temporary situations to existing buildings.

As the scope of thermal comfort for a conditioned house differs from that for a free running house, methods and criteria to determine thermal acceptability should be determined in relation to the house operation mode. ISSO, 2004 relies on the Fanger model (PMV) and other studies based on the indoor operative temperature as a function of mean monthly outdoor air temperature.

However, this method is limited to only three categories, and does not differentiate between buildings with thermal performance below 65% thermal acceptance. Moreover, it has been developed for office buildings, and its applicability to residential buildings needs to be examined further. These restrictions make it inapplicable for the purpose of evaluating house performance for a house rating scheme.

The thermal performance of buildings can also be evaluated by employing the "degree hour" method. A degree hour is the amount of time spent above or below a standard reference thermal comfort zone during an hour. This method is used to express the length of time and how far the indoor temperature falls below or above the comfort temperature. It can be used to evaluate or predict a building's performance and to estimate the annual energy requirement of buildings (Buyukalaca et al., 2001; De Dear and Hart, 2002; Christenson et al., 2005).

An important issue in employing this method for the thermal performance evaluation of buildings is the method of summing up the length of time that the comfort range is exceeded, because the value of different degree hours is not the same. For this purpose, weighting factors are proposed by some sources (International Standards Organisation, 2003; ISSO, 1990; Olesen, 2004), and in the GBA (Government Buildings Agency) in the Netherlands, as described in van der Linden (van der Linden et al., 2002). But these are only applicable for conditioned buildings because they propose a weighting factor which depends on Fanger's (PMV) model[3], the inapplicability of which to free running houses has already been discussed.

3.2.1 Computing Degree Hours for Free Running Houses

The weighting factor for computing "degree hours" in free running houses can be determined on the basis of the percentage of dissatisfied people in naturally ventilated houses for each discomfort hour. However, extensive research has still not produced a framework or model to determine the PPD for naturally ventilated buildings.

A study in South Brazil (Xavier and Lamberts, 2001) showed a probit regression of dissatisfied people in a number of naturally ventilated schools. It showed the percentage of dissatisfied occupants when the temperature changed from the comfort temperature range. This method can be applied for office buildings as well as schools, but not for residential buildings, because there are inflexible conditions for occupants in both the former, unlike in dwellings. As noted above, the behaviour of residents and their use of clothing is not predictable in residential buildings.

In a study to assess the thermal performance of free running houses, Willrath (1998) used an equivalence between degree hours of discomfort, in which ten degree

---

[3] Based on ISO 2003, the time during which the PMV exceeds the comfort boundaries is weighted with a factor which is a function of the PPD on a yearly basis, and is expressed as follows:

$$Wf = PPD_{actual\ PMV}/PPD_{PMV\ limit} \qquad (3.11)$$

Where $PPD_{actual\ PMV}$ is the instantaneous value in which the PPD exceeds the limit $PPD_{PMVlimit}$ which depends on the class of comfort. The warm period is calculated from $\sum Wf * time\ hours$, where $PMV > PMV_{limit}$ and the cold period is obtained from it when $PMV < PMV_{limit}$. The entirety of the resulting "weighting factor * time" is named "weighting time" in hours, which is applicable for the assessment of long term conditions but not for free running buildings. A similar method has been introduced in ISSO for the sum of weighted temperatures exceeding hours (Wf), as explained by Breesch and Janssens (2004). Wf is considered directly as a proportion of PPD, in which an hour with 20% dissatisfied occupancy counts twice as much as an hour with 10% dissatisfaction. Based on van der Linden et al. (2002) the GBA (Government Buildings Agency) in the Netherlands also applied Fanger's PPD as a criterion for calculating the extent of excess temperatures. Over the period in which it exceeded a $PMV = 0.5$ (PPD = 10%), a weighting directly proportional to the PPD was applied, which means 1 h with 20% dissatisfied people was weighted twice as much as 1 h with 10% dissatisfied occupants.

hours of discomfort were equivalent to 1 h at ten degrees of discomfort, or 10 h at one degree over or under the comfort limit. For different occupied zones, a weighting in proportion to its area as a fraction of the total was given to the total degree hours of each zone.

The above equivalence between degree hours is a simple concept that can be applied to evaluate the free running performance of a house on the basis of thermal comfort. Although there is no linear relationship between the percentage of dissatisfied people and the degree range, as there is no framework or model developed for discomfort beyond the comfort zone, the above simplified method of degree hours appears to be suitable for evaluating a building's free running performance.

The main criteria for evaluating free running buildings as a basis for a comfort based rating scheme can be set to:

- the boundaries of the comfort zone for defining thermal neutrality
- the temperature exceedance degree hour method.

3.3 Indicators to Measure the Thermal Performance of Houses for Rating Purposes

3.3.1 Conditioned Mode

Conditioned houses have their thermal performance evaluated in terms of energy, which is an aggregation of the heating and cooling energy (sensible + latent) required for maintaining comfort temperatures within particular zones for specifically nominated time periods. This is a method applied in nearly all energy based rating systems. Annual energy required is expressed in $MJ/m^2.annum$ as an indicator of house thermal performance in the conditioned operation mode.

Despite the unreliability of this indicator for the purpose of a rating scheme, as previously discussed, we will employ it in order to make a comparison between the current rating scheme and a rating scheme that is proposed in this book.

3.3.2 Free Running Mode

Houses in the free running operation mode have their thermal performance evaluated in terms of annual Degree Discomfort Hours (DDH). This is calculated from a combination of "heating and cooling discomfort hours". Heating energy requirements for a conditioned building and "heating" discomfort hours for the building in free running mode are indicators of a winter building performance. Likewise cooling energy requirements and "cooling" discomfort hours have been determined to investigate summer performance. Figure 3.4 shows the state of these two categories associated with the boundaries of the comfort zone.

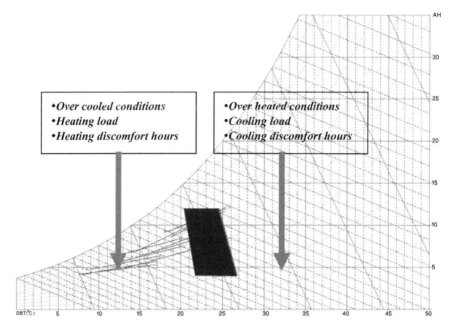

Fig. 3.4 The state of discomfort hours relative to the comfort zone

3.3.3 How an Indicator Points to Building Efficiency

The efficiency of an evaluated building can be determined by the value of a specified indicator in a building performance evaluation system. The value of indicators can be determined by its numerical value. It should be noted that a greater number does not always point to the building being more efficient. For instance, if the indicator for evaluating a building's energy performance is the amount of energy required or energy used (MJ/m^2), obviously a building with less energy required would be the most energy efficient building. However, in building rating systems efficiency is indicated by giving a grade of stars or other values such as golden, silver and bronze, or different colours. In that system, each grade is an indication of a range of numbers. Such a system must determine the value of the indicators used for grading buildings.

3.3.3.1 Correlation Between Indicators of Building Efficiency

It appears that the two indicators mentioned above to evaluate the thermal performance of houses are strongly related to each other, particularly when DDHs are adjusted with area weighting. Occupants are basically more willing to turn on air conditioning or to load energy for space heating and cooling when the indoor environment is not thermally comfortable. Thus there may not be a significant difference between evaluating the thermal performance of houses in different operation modes. However, even if there is a strong correlation between those two indicators, it does not mean that an efficient architectural design for a free running house would be similar to an efficient architectural design for that house in conditioned mode. Therefore the value of an efficient free running house could be underestimated if it is to be evaluated on the basis of its thermal performance in conditioned operation mode. This supposition will be tested and discussed further in the next chapters.

3.4 Summary

The principal work on comfort by Fanger, which was based on the heat balance between the surrounding environment and subjects in steady state conditions, has been replaced by recent studies, and adaptive thermal models such as those produced by Humphries, Auliciems and de Dear. These take into account the effect of acclimatization. This makes the adaptive comfort model applicable for naturally ventilated buildings. The de Dear comfort equation, which is expressed in ASHRAE 55 (2004), has been used in this book as a basis for calculating *Degree Discomfort Hours* in free running dwellings.

Annual Degree Discomfort Hours (DDH) can be used as an indicator to evaluate the thermal performance of a free running house, and a normalized energy load (MJ/m^2.annum) as an indicator for the evaluation of the thermal performance of houses in the conditioned operation mode. Although the latter is not an ideal indicator for the evaluation of the energy performance of houses, we will employ it as common indicator that is used in the majority of energy based rating systems, in order to examine its applicability for addressing aspects of efficiency in architectural design, and to compare the current energy based rating schemes with ones based on thermal comfort.

It should be noted that the concept of free running houses in this book refers to naturally ventilated houses without any mechanical equipment to improve their indoor thermal condition. This definition is not applicable in warm humid climates in which the benefits of using fans for improving summer performance would not be ignored. However, the definition used here is appropriate for a moderate climate, such as that of Sydney in Australia, in which the effect of the "intelligent management of buildings" by occupants should not be ignored. The means for this includes controlling shutters and blinds to take advantage of outside weather, particularly with regard to directing natural ventilation into the inside environment.

References

ASHRAE: ASHRAE Handbook – 1981 Fundamentals. Atlanta: American Society of Heating, Refrigeration and Air Conditioning Engineers, Inc. (1981)

ASHRAE: ASHRAE Handbook – 2001 Fundamentals. American Society of Heating. Refrigeration and Air Conditioning Engineers, Inc., Atlanta, GA (2001)

ASHRAE: Handbook: Heating Ventilating and Air-Conditioning Applications. Inch-Pound Edition (2003)

ASHRAE: ASNI/ASHRAE Standard 55-2004, Thermal Environmental Conditions for Human Occupancy. Atlanta: American Society of Heating, Refrigeration and Air Conditioning Engineers, Inc. (2004)

Auliciems, A.: Towards a psycho-physiological model of thermal perception. Int. J. Biometeorol. **25**(2), 109–122 (1981)

Auliciems, A., de Dear, R.: Air Conditioning in Australia I – human thermal factors. Architectural Science (1986)

Auliciems, A., Szokolay, S.V.: Thermal Comfort. Brisbane, QLD: PLEA in association with Department of Architecture, University of Queensland (1997)

Azer, N.Z., Hsu, S.: The prediction of thermal sensation from a simple model of human physiological regulatory response. ASHRAE Trans. **83**(Pt 1) (1977)

Baker, N., Standeven, M.: Thermal comfort for free running buildings. Energy Build. **23**(3), 175–182 (1996)

Benzinger, T.H.: The physiological basis for thermal comfort. First International Indoor Climate Symposium, Copenhagen: Danish Building Research Institute, pp. 441–476 (1979)

Berglund, L.: Mathematical models for predicting the thermal comfort response of building occupants. ASHRAE Trans. **84**, 735–749 (1978)

Bouden, C., Ghrab, N.: An adaptive thermal comfort model for the Tunisian context: field study results. Energy Build. **37**(9), 952–963 (2005)

Brager, G.S., de Dear, R.J.: Thermal adaptation in the built environment: a literature review. Energy Build. **27**(1), 83–96 (1998)

Brager, G.S., de Dear, R.: A standard for natural ventilation. ASHRAE J. **42**(10), 21–28 (2000)

Brager, G.S., de Dear, R.: Climate, comfort and natural ventilation: a new adaptive comfort standard for ASHRAE standard 55. In: Conference Proceedings: Moving Thermal Comfort Standards into the 21st Century, pp. 60–77. Oxford Brookes University, Windsor (2001)

Breesch, H., Janssens, A.: Uncertainty and Sensitivity Analysis of the Performances of Natural Night Ventilation. In: Proceedings of the 9th International Conference on Air Distribution in Rooms, Coimba, Portugal. http://hdl.handle.net/1854/2713 (2004). Accessed 17 Sep 2008

Buyukalaca, O., Bulut, H., Yilmaz, T.: Analysis of variable-base heating and cooling degree-days for Turkey. Appl. Energy. **69**(4), 269–283 (2001)

Chappells, H., Shove, E.: Comfort Paradigms and Practices in Future Comforts: Re-conditioning the Urban Environment, Workshop. London: The Policy Studies Institute (2004)

Christenson, M., Manz, H., Gyalistras, D.: Climate warming impact on degree-days and building energy demand in Switzerland. Energy Conversion and Manag. **47**(6), 671–686 (2005)

Chung, T.M., Tong, W.C.: Thermal comfort study of young Chinese people in Hong Kong. Build. Environ. **25**(4), 317–328 (1990)

Davis Energy Group: Comfort Reports (No. P500-04-009-A4). California: California Energy Commission (2004)

de Dear, R.: Thermal comfort in practice. Indoor Air **14**(7), 32–39 (2004)

de Dear, R.J., Brager, G.S.: An adaptive model of thermal comfort and preference. ASHRAE Trans. **104**(1a), 145–167 (1998)

de Dear, R., Brager, G.S.: The adaptive model of thermal comfort and energy conservation in the built environment. Int. J. Biometeorol. **45**(2), 100–108 (2001)

de Dear, R.J., Brager, G.S.: Thermal comfort in naturally ventilated buildings: revisions to ASHRAE standard 55. Energy Build. **34**(6), 549–561 (2002)

de Dear, R., Brager, G.S., Cooper, D.: Developing an Adaptive Model of Thermal Comfort and Preference (Final Report on ASHRAE RP-884). Sydney: MRL (1997)

de Dear, R.J., Fountain, M.E., Popovic, S., Watkins, S., Brager, G., Arens, E. et al.: A Field Study of Occupants Comfort and Office Environment in a Hot- humid Climate (Final report ASHRAE RP-702). Sydney: Macquarie University (1993)

de Dear, R., Hart, M.: Appliance Electricity End-Use: Weather and Climate Sensitivity. Sydney: Division of environmental and life sciences, Macquarie University (2002)

Ellis, F.P.: Thermal comfort in warm and humid atmospheres; observations on groups and individual in Singapore. J. Hyg. (Lond). **51**(3), 386–404 (1953)

References

Fanger, P.O.: Calculation of thermal comfort: introduction of a basic comfort equation. ASHRAE Trans. **73**(part 2), III.41–III.44.20 (1967)

Fanger, P.O.: Thermal Comfort. Danish Technical Press, Copenhagen (1970)

Fanger, P.O., Toftum, J.: Extension of the PMV model to non-air-conditioned buildings in warm climates. Energy Build. **34**(6), 533–536 (2002)

Feriadi, H., Wong, N.H.: Thermal comfort for naturally ventilated houses in Indonesia. Energy Build. **36**(7), 614–626 (2004)

Forwood, G.: What is thermal comfort in a naturally ventilated building? In: Nicol, F., Humphreys, M., Skes, O., Roaf, S. (eds.) Standards for Thermal Comfort, Indoor Air Temperature Standards for the 21st Century, pp. 176–181. E & FN Spon, London (1995)

Fountain, M., Brager, G., de Dear, R.: Expectations of indoor climate control. Energy Build. **24**(3), 179–182 (1996)

Gagge, A.P.: Rational temperature indices of mans thermal environment and their use with a 2-nude model of his temperature relation, Federation Proceedings. pp. 1572–1582 (1973)

Gagge, A.P., Herrington, L.P., Winslow, C.-E.A.: The thermal interchanges between the human body and its atmospheric environment. Am. J. Hyg. **26**, 84–102 (1937)

Gagge, A.P., Stolwijk, J.A., Hardy, J.D.: Comfort and thermal sensations and associated physiological responses at various ambient temperatures. Environ. Res. **1**, 1–20 (1967)

Gagge, A.P., Stolwijk, J.A.J., Nishi, Y.: An effective temperature scale based on a simple model of human physiological regulatory response. ASHRAE Trans. **77**, 247–262 (1971)

Givoni, B.: Man, Climate and Architecture. Applied Science Publishers Ltd, London (1976)

Givoni, B.: Climate Considerations in Building and Urban Design. Van Nostrand Reinhold, New York, NY (1998)

Grivel, F., Candas, V.: Ambient temperatures preferred by young European males and females at rest. Ergonomics. **34**, 365–378 (1991)

Hensen, J.M.: Literature review on thermal comfort in transient conditions. Build. Environ. **25**(4), 309–316 (1990)

Holm, D., Engelbrecht, F.A.: Practical choice of thermal comfort scale and range in naturally ventilated buildings in South Africa. J. S. Afr. Inst. Civ. Eng. **47**(2), 9–14 (2005)

Humphreys, M.A.: Field Studies of Thermal Comfort Compared and Applied (BRE-CP 76/75). Building Research Establishment, Garston (1975)

Humphreys, M.A.: Field studies of thermal comfort compared and applied. Build. Serv. Eng. **44**, 5–27 (1976)

Humphreys, M.A.: Outdoor temperature and comfort indoors. Build. Serv. Eng. **6**(2), 92–105 (1978)

Humphreys, M.A., Fergus Nicol, J.: The validity of ISO-PMV for predicting comfort votes in every-day thermal environments. Energy Build. **34**(6), 667–684 (2002)

Humphreys, M.A., Nicol, J.F.: An adaptive guideline for UK office temperatures. In: Nicol, F., Humphreys, M., Skes, O., Roaf, S. (eds.) Standards for Thermal Comfort, Indoor Air Temperature Standards for the 21st Century. E & FN Spon, London (1995)

Humphreys, M., Nicol, J.F.: Understanding the adaptive approach to thermal comfort. ASHRAE Tech. Data Bull. **14**(1), 1–14 (1998)

Ihab, M.K.E.: Designing for indoor comfort: a system model for assessing occupant comfort in sustainable office buildings, Proceedings of the Solar 2002 Conference, Reno, Nevada. American solar energy society, American Institute of Architects Committee on the Environment, pp. 485–494 (2002)

International Standards Organisation: ISO/DIS 7730 Ergonomics of the Thermal Environment-Analytical Determination and Interpretation of Thermal Comfort Using Calculation of the PMV and PPD Indices and Local Thermal Comfort: International Standards Organisation (2003)

ISO 7730: Moderate Thermal Environment – Determination of the PMV and PPD Indices and Specifications for Thermal Comfort. Geneva, Switzerland: International Organisation for Standardisation (1995)

ISSO: Design of Indoor Conditions and Good Thermal Comfort in Buildings (in Dutch) (No. ISSO Research Report 5). Netherlands (1990)

ISSO: Thermal Comfort as Performance (No. ISSO Research Report 58.2). Rotterdam. Netherlands (2004)

Jones, B.W.: Capabilities and limitations of thermal models. Conference Proceedings: Moving Thermal Comfort Standards into the 21st century, Cumberland Lodge, Windsor, UK. Oxford Brookes University, pp. 112–121 (2001)

Karynono, T.H.: Thermal comfort in the tropical south-east Asian region. Archit. Sci. Rev. **39**(3), 135–139 (1996)

La Gennusa, M., Nucara, A., Rizzo, G., Scaccianoce, G.: The calculation of the mean radiant temperature of a subject exposed to solar radiation–a generalised algorithm. Build. Environ. **40**(3), 367–375 (2005)

Lee, J.-Y., Choi, J.-W.: Influences of clothing types on metabolic, thermal and subjective responses in a cool environment. J. Therm. Biol. **29**(4–5), 221–229 (2004)

Lovins, A.: Air Conditioning Comfort: Behavioural and Cultural Issues: E Source, Inc., Boulder, Colorado. Available from: http://www.osti.gov/energycitations/product.biblio.jsp?osti_id=55564 (1992)

Lyons, P.: Window Performance for Human Thermal Comfort (Final report to the national fenestration rating council). Melbourne: Centre for the Built Environment (2006)

Lyons, P., Arasteh, D., Huizenga, C.: Window performance for human thermal comfort. ASHRAE Trans. **73**(2), 4.0–4.20 (1999)

Mayer, E.: Objective criteria for thermal comfort. Build. Environ. **28**(4), 399–403 (1993)

McNall, P.E.J., Jaax, J., Rohles, F.H., Nevins, R.G., Springer, W.: Thermal comfort (thermally neutral) condition or three levels of activity. ASHRAE Trans. **73**(part 1), 1.3.1.–1.3.13 (1967)

Melikov, A.K.: Air Movement at the Neck of the Human Body. Proceedings of Indoor Air, Nagoya, Japan, pp. 209–214 (1996)

Melikov, A.K.: Personalized ventilation. Indoor Air. **14**(7), 157–167 (2004)

Nakano, J., Tanabe, S., Kimura, K.: Differences in perception of indoor environment between Japanese and non-Japanese workers. Energy Build. **34**(6), 615–621 (2002)

Nicol, F., Aulicien, A.: A Survey of Thermal Comfort in Pakistan: Toward New Indoor Temperature Standards (Final Report to the Overseas Development) UK: Administration. Oxford Brookes University, School of Architecture (1994)

Nicol, J.F., Raja, I.A., Allaudin, A., Jamy, G.N.: Climatic variations in comfortable temperatures: the Pakistan projects. Energy Build. **30**(3), 261–279 (1999)

Nicol, F., Roaf, S.: Pioneering new indoor temperature standards: the Pakistan project. Energy Build. **23**(3), 169–174 (1996)

O'Callaghan, P.W.: Building for Energy Conservation. Pergamon Press, New York, NY, Oxford (1978)

Olesen, B.W.: International standards for the indoor environment. Indoor Air. **14**(supplement 7), 18–26 (2004)

Olesen, B.W., Seppanen, O., Boerstra, A.: Criteria for the Indoor Environment for Energy Performance of Buildings: A New European Standard, Windsor Conference. http://nceub.org.uk/uploads/Olesen.pdf (2006). Accessed 21 Sep 2008

Olgyay, V.: Design with Climate: Bioclimatic Approach to Architectural Regionalism. Princeton University Press, Princeton, NJ (1963)

Parsons, K.C.: The effects of physical disability, gender, acclimation state and the opportunity to adjust clothing, on requirements for thermal comfort, Conference Proceedings: Moving Thermal Comfort Standards into the 21st century, Cumberland Lodge, Windsor, UK. Oxford Brookes University (2001)

Parsons, K.C.: Human Thermal Environments. Taylor and Francis, London, New York, NY (2003)

Szokolay, S.V.: Environmental Science Hand Book for Architects and Builders. The Construction Press, London (1980)

Szokolay, S.V.: Introduction to Architectural Science: The Basis of Sustainable Design. Architectural Press, Oxford (2004)

References

van der Linden, K., Boerstra, A.C., Raue, A.K., Kurvers, S.R.: Thermal indoor climate building performance characterized by human comfort response. Energy Build. **34**(7), 737–744 (2002)

van der Linden, A.C., Boerstra, A.C., Raue, A.K., Kurvers, S.R., de Dear, R.J.: Adaptive temperature limits: A new guideline in the Netherlands: A new approach for the assessment of building performance with respect to thermal indoor climate. Energy Build. **38**(1), 8–17 (2006)

Watt, J.R.: Evaporative Air Conditioning. The Industrial Press, New York, NY (1963)

Williamson, T.J., Coldicutt, S., Riordan, P.: Comfort preferences or design data? In: Nicol, F., Humphreys, M., Skes, O., Roaf, S. (eds.) Standards for Thermal Comfort, Indoor Air Temperature Standards for the 21st Century, pp. 50–58. E & FN Spon, London (1995)

Willrath, H.: The Thermal Performance of Houses in Australian Climates. Unpublished PhD thesis, University of Queensland, Brisbane (1998)

Wong, N.H., Feriadi, H., Lim, P.Y., Tham, K.W., Sekhar, C., Cheong, K.W.: Thermal comfort evaluation of naturally ventilated public housing in Singapore. Build. Environ. **37**(12), 1267–1277 (2002)

Xavier, A.A.P., Lamberts, R.: Thermal comfort zones for conditioned and free running buildings in Florianopolis, South Brazil. Conference Proceedings Moving Thermal comfort standard into the 21st century, Cumberland Lodge, Windsor, UK. Oxford Brookes University, pp. 235–245 (2001)

Yaglou, C.P.: The comfort zone for men at rest and stripped to the waist. Trans. ASHVE. **33**, 165–179 (1927)

Yoshida, J.A., Nomura, M., Mikami, K., Hachisu, H.: Thermal comfort of severely handicapped children in nursery schools in Japan. Proceedings of the IEA 2000/HFES Congress, San Diego, CA, USA, pp. 712–715 (2000)

Young, A.J.: Effects of aging on human cold tolerance. Exp. Aging Res. **17**, 205–213 (1991)

Young, A.J., Lee, D.T.: Aging and human cold tolerance. Exp. Aging Res. **23**, 45–67 (1997)

Chapter 4
Modelling Efficient Building Design: Efficiency for Low Energy or No Energy?

As described in the previous chapters, HRS have been developed to promote energy efficient design and to consequently reduce energy requirements in the buildings sector. However, zero energy buildings and passive design buildings, whose thermal performance should be evaluated in their free running operation mode, are not a target in current HRS. The main purpose of this book is to recommend the evaluation of the free running performance of buildings on the basis of thermal comfort instead of energy consumption. A major question is whether an energy efficient building is one that demonstrates efficient performance in the free running operation mode. If the answer is yes, then any attempts at making a building energy efficient would produce improved thermal performance of that building in free running operation. If the answer is no, then devices and regulations for free running buildings should differ from those for conditioned buildings, leading to the need to revise current energy based rating schemes.

This chapter aims to illustrate the similarities and differences between the performances of a building in different operation modes in response to changed design features. This will demonstrate that an efficient energy based building is not always an efficient free running building. The recognition of this result should therefore be considered in the development of the next generation of HRS.

The chapter is divided into two sections: the first section describes the tools, criteria and methods employed for building performances evaluation and then applies these to a number of house samples in order to evaluate their thermal performances in two different operation modes. The second section reports the results of the thermal performance evaluations of the sample houses in response to changed design features, and then investigates the relationship between the thermal performances of those houses in different operation modes.

4.1 Building Performance Evaluation

Building performance assessment is an approach to the design and construction of a building which systematically compares the actual or expected performance of buildings, in order to explicitly document criteria for their expected performance (Preiser, 2005; Preiser and Vischer, 2005).

Building performance is assessed by a numerical measure of an indicator. This is usually expressed in terms of annual energy requirements, thermal comfort, embodied energy, cost effectiveness, environmental impact and other parameters, depending on the purpose of the evaluation. The indicator is a value derived from a parameter that describes the state of a building. Thus, for example, for thermal performance, two different indicators can be defined to evaluate the building's performance, depending on its state (conditioned or free running). The thermal quality of a building can be evaluated in terms of annual energy requirements in its conditioned mode, or an aggregated annual thermal comfort condition in its free running mode. The latter demonstrates the actual performance of the building, and addresses multiple aspects of efficiency in a particular architectural design (Kordjamshidi et al., 2005a).

Building performance assessment is used for post-occupancy performance evaluation to enable improvements to be made in further building construction or renovation (Bordass and Leaman, 2005), often by using simulation programs. The methods used for building performance evaluation are:

- calculation
- experimentation
- simulations.

The first of these methods, calculation, was done manually in earlier times by architects and building services engineers, using pre-selected design conditions, and they often resorted to the "rule-of-thumb" methods of estimation to extend conventional design concepts. This approach frequently led to poor assessment of energy performance, because of excessive part-load operations (Hong et al., 2000). The "bin method", as a simple hand-calculation procedure, has also been used for calculating energy requirements in buildings based on the assumption of steady state conditions and simple building descriptions, but this method is also limited in its applicability, in addition to being unreliable (Hanby, 1995; Klein, 1983).

There are also problems with physical experimentation, in that it is often expensive and time consuming. Besides, it is difficult to modify parameters to determine the effect of modification on the thermal performance of a building. This method therefore is not suitable at the design stage, but could be appropriate for evaluating the reliability of any result from a study which employed the other two methods.

The third general approach, simulation, however, has been highly developed with the advancements in computer technology, and many simulation models now exist to predict the thermal behaviour of buildings. Simulation computer programs are flexible, accurate and reliable tools for designing and analysing the efficiency of a building design. These advantages have been confirmed by a number of reviews of simulation models (Al-Homoud, 2000; Clarke, 2001; Hong et al., 2000; Littler, 1982; Sowell and Hittle, 1995). An appropriate computer simulation can provide information on thermal performance that is as accurate as a physical experiment, while involving less time and expenditure. In the design stage for a building, computer tools such as full-scale mock-ups and simulations of interior and

4.1 Building Performance Evaluation 55

exterior spaces can provide further information for assessing all the different aspects of design. Such programs are thus the most appropriate for the evaluation of the thermal performance of houses at the design stage.

The simulation method is therefore employed here for the purpose of evaluating the thermal performance of houses in different operation modes. In addition to the above-mentioned advantages, there are two further important reasons for this. The first is that it helps to evaluate the thermal performance of a large number of houses within a limited time, and the second is that it is the most common method used to evaluate the thermal performance of buildings in HERS. Since the development of HERS is the core of this book, any improvements based on this method should be applicable to these rating systems.

4.1.1 Building Simulation Programs

Simulation programs are undergoing continuous development. Most building simulations perform hour-by-hour calculations for analysis, and all use algorithms and models which provide approximate representations of the heat transfer mechanism of the physical elements to the environment, to other buildings and to internal energy resources. A number of developed programs are listed in the Building Energy Software Tools Directory,[1] which categorises tools on the basis of subject, platform and country. The subject category is classified as follows:

- Whole Building Analysis
  - energy simulation
  - load calculation
  - renewable energy
  - retrofit analysis
  - sustainability/green building
- Code and Standards
- Materials, Components, Equipment, and system
- Other applications

Under the energy simulation category there are about 100 software programs, mainly developed in the US, each of which has a particular weakness or limitation, depending on the purpose for which it is used.

Software developed in Australia for HERS has been chosen here for simulating the thermal performance of samples. A review of the software packages developed in Australia is presented in the following for general information.

---

[1] See: http://www.eere.energy.gov/buildings/tools_directory/

4.1.2 Criteria for Modeling the Thermal Performance of Buildings in Two Different Operation Modes

A comparison between the thermal performances of buildings in similar conditions and using constant criteria is made in the following. *Similar conditions* refer to both climate and simulation programs, and *constant criteria* refer to features such as thermal comfort conditions, occupancy scenarios and thermostat settings. For the modeling and evaluation of the thermal performances of a building the following aspects are covered:

- Samples (or case studies)
- Simulation program
- Climate
- Thermal comfort conditions
  - Comfort condition boundaries for free running buildings
  - Thermostat settings for buildings in conditioned operation mode
- Occupancy scenarios
- Indicators

4.1.2.1 Samples

The sample of typical houses developed initially by SOLARCH (2000) was selected for examination in this research. Typical houses are representative of popular practice in national architectural design; therefore the result of an examination of these samples reflects the performances of representative buildings of the entire nation and can then be generalized for a larger group of buildings, from which subpopulations of interest can be extrapolated. In the SOLARCH study, to establish typical houses, 100 designed houses were collected from Project Home Display Villages throughout Sydney. Six typical houses were identified from these samples to represent the following parameters:

- single and double storey dwellings
- large and small dwellings
- dwellings with potential for adaptation of plans for a range of ideal orientations

These six house designs selected differ in size and planning and cover a broad range of those available in Sydney. The floor plans of the samples are shown in Figs. 4.2, 4.3, 4.4, 4.5, 4.6, and 4.7. General metric descriptions of the samples are presented in Table 4.1. Since window areas in different orientations differ among these six typical houses, Fig. 4.1 represents this variation. It shows the percentage of window-to-wall ratio for each orientation of each typical house. The ratio is computed from [window area/total wall area], in which total wall area means (window + wall) area. The full description of the components of these two types of construction as used in this book is shown in Table 4.2.

4.1 Building Performance Evaluation

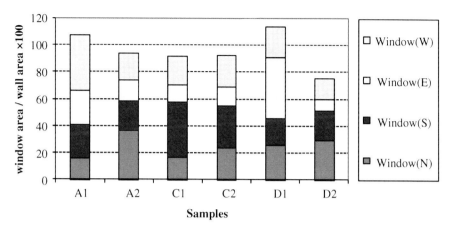

Fig. 4.1 The percentage of [window/(window + wall)] for each orientation of typical houses

Fig. 4.2 Plan of typical house (House A1)

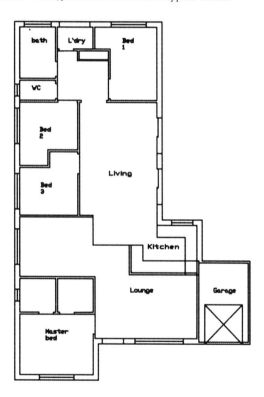

4.1.2.2 Simulation Programs

A simulation program developed in Australia for the purpose of HRS was used for modeling samples. A summary of developed building simulation programs in Australia is given below as background.

Fig. 4.3 Plan of typical house (House C1)

Building Simulation Programs in Australia

Simulation programs for buildings started to be developed in Australia in the 1970s, with programs such as TEMPER, BUNYIP, Star performer, and CHEETAH. The latter is a development of the program ZSTEP3 (Delsante, 1987) and is well documented and validated (Delsante, 1995b). According to Clarke (2006) it is "considered by many to be equal to any in the world". This program calculates hourly temperatures, and the heating and cooling energy requirements of small buildings. It is based on the total–zone response factor method, which uses measured hourly weather data (temperature, solar irradiance and wind speed).

The first developed computer program for the establishment of house ratings was based on CHEETAH (Ballinger and Cassell, 1994), the core energy software model developed by the CSIRO for Australian climates (Ballinger, 1998a). Most modelling systems used in HERS, such as NatHERS, FirstRate and QuickRate, BERS, QRate and ACTHERS are based on this engine. NatHERS and BERS simulate the operational energy used in a home, while FirstRate, QRate, ACTHERS and QuickRate are correlation programs, which do not carry out simulations. A study by CSIRO has found that BERS and NatHERS software give similar results for a range of house variations. They produce the same results because they use the same engine, CHEENATH. The former, BERS, originally just contained star rating settings for 12

4.1 Building Performance Evaluation

Fig. 4.4 Plan of typical house (House D1)

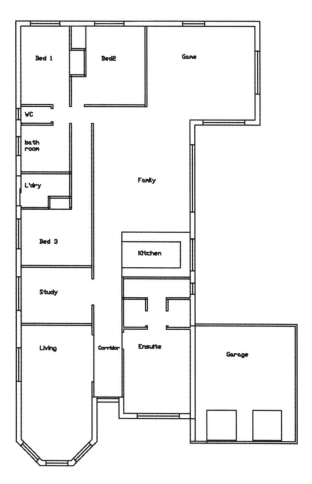

climate zones in Queensland, but later came to cover all climate zones of Australia (Ballinger, 1998a).

NatHERS simulation software has been accepted nationally as the benchmark tool for assessing the thermal performance of houses, and is known to be a tool unique to Australia in addressing the issue of energy efficiency. It allows houses to be rated on the basis of energy consumption and has been used for several years in both regulatory and voluntary rating schemes. However, NatHERS has had a number of shortcomings, which led to criticism of the software as the foundation of a national regulatory rating scheme and as an appropriate tool for efficient design analysis. The main shortcomings are that:

- it ignores the physiological effect of natural ventilation in its computation for estimating energy requirements;
- it is limited to only three conditioned zones in a house for simulation.

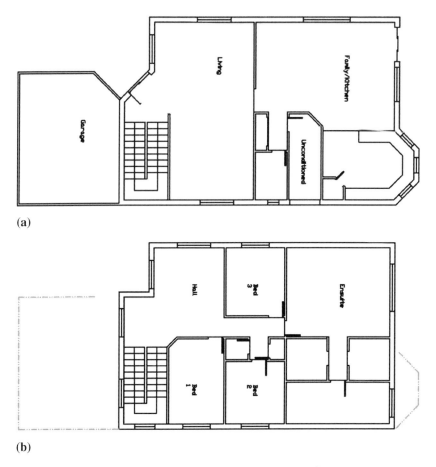

Fig. 4.5 Plan of typical house (House A2). (**a**) Ground floor. (**b**) First floor

A new rating tool, AccuRate, which is a development of NatHERS, was created in 2005 to address these shortcomings (Isaacs, 2005). Its engine is an enhanced version of the CHEENATH simulation engine. Its ventilation model was completely revised to include the physiological cooling effect of ventilation in computing the cooling energy requirement for conditioned houses. The process of this inclusion is described by Delsante (2005) and is briefly discussed further under Sect. 4.1.2.4. The effect of that improvement has been demonstrated in a comparative study of NatHERS and AccuRate (Isaacs, 2004). This study demonstrated a substantial reduction in predicting the cooling energy requirement of a well-ventilated house for seven different climates. The other aspects which are improved in AccuRate are:

- room by room zoning (up to 99 zones)
- flexibility for making a particular construction
- improvement in the modelling of reflective insulation

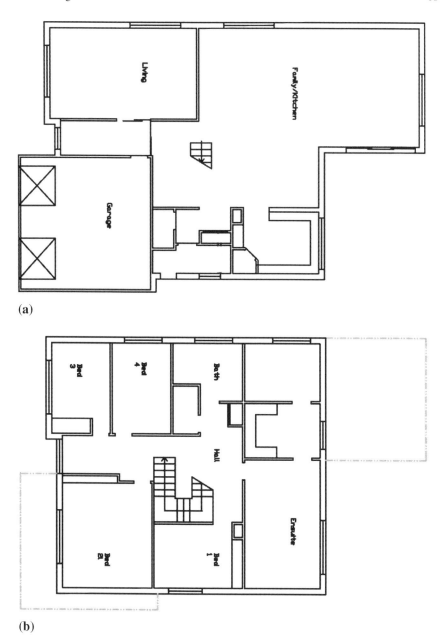

Fig. 4.6 Plan of typical house (House C2). (**a**) Ground floor. (**b**) First floor

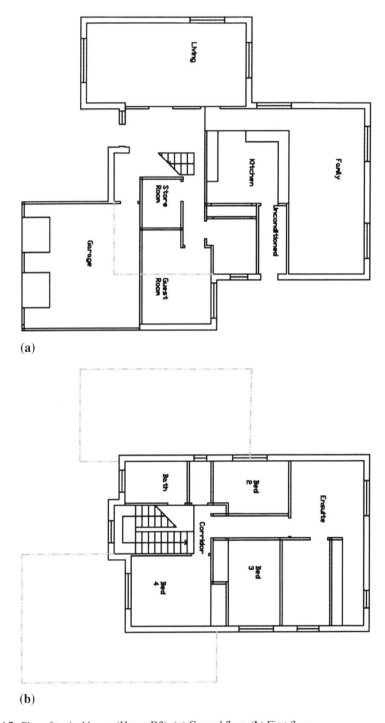

Fig. 4.7 Plan of typical house (House D2). (**a**) Ground floor. (**b**) First floor

4.1 Building Performance Evaluation

Table 4.1 Window, wall, ceiling, floor areas of typical houses

House	Number of floors	External wall area (m^2)	Window area (m^2)	Ceiling area (m^2)	Internal wall (m^2)	Floor area (m^2)
1A	1	137	32.4	138.2	96.6	138.2
1C	1	150	24.8	155.4	88.1	155.4
1D	1	196.5	45.9	244.9	160.4	244.9
2A	2	256.7	50	166	156.1	292.8
2C	2	260	56.5	136.3	182.3	315.7
2D	2	234	40	144.4	174.4	229

Table 4.2 General description of the materials used in the simulation for the 6 base houses

Construction	Heavy weight (HW)	Light weight (LW)
External wall	Brick veneer (uninsulated)	Weatherboard (uninsulated)
	External colour: light	External colour: light
	Internal colour: Medium	Internal colour: Medium
Internal wall	Plasterboard on studs	Plasterboard on studs
	Colour: (Medium)	Colour: (Medium)
Window	Single glassed clear (4 mm)	Single glassed clear (4 mm)
Window internal cover	Closed weave	Closed weave
Window external cover	None	None
Window frame	Aluminium	Timber
	Colour: (Medium)	Colour: (Medium)
Door	Timber (hollow)	Timber (hollow)
	External (solid)	External (solid)
Roof	Roofing tiles	Roofing tiles
	Colour: (Medium)	Colour: (Medium)
Ceiling	Plasterboard 13 mm	Plasterboard 13 mm

- user interface improvement
- detailed hourly output with adjustment for floor area correction[2]

Although rating houses by AccuRate would appear to be more precise than using NatHERS, the shortcomings in AccuRate are still considerable, namely the lack of capability to rate free running buildings completely, and inflexibility in dealing with different occupancy scenarios.

Notably, AccuRate software is not flexible enough to deal with variations in occupant behaviour and in modelling occupant interactions. Any condition in which doors, windows and blinds need to be opened and closed for specific purposes cannot be simulated by this software. A model has been set into the software, by which

---

[2]Small houses compared to large houses usually are penalized in the star rating scheme because of the basis of rating (MJ/m^2), which was discussed in Sect. 2.4.3. AccuRate has addressed this concern by developing an area correction factor (Isaacs, 2005).

doors and windows are opened when the indoor environmental temperature exceeds the boundaries of thermal comfort, and outdoor temperature is less than indoor temperature. The software in this situation opens the windows of a simulated house in order to reduce indoor temperature by increasing heat transfer and natural ventilation, without considering whether the house is occupied or not. It is obvious that if a house is not occupied, windows cannot be opened automatically; however, this reality is ignored in the design of the AccuRate software. Another issue in the setting of the software program is that the state of occupation in relation to the operation of blinds and external shading is ignored. This simulation therefore does not mirror the actual performance of a house with real occupants.

Nevertheless, the shortcomings in the AccuRate and NatHERS software are not as important as they might appear. The main limitation of the inability of the software to deal with various occupancy scenarios is a problem that also exists in all the other available software programs. The few software packages elsewhere which have attempted to deal with this problem, such as Energy Plus and Energy10, appear to be difficult and complex in scheduling the opening of windows and doors in a manner compatible with a rating scheme.

There are other advantages to using AccuRate software in addition to those mentioned above. Both software programs, NatHERS and AccuRate, have been designed for the Australian climate and have the capability of analysing the energy consumption and hourly indoor temperature for non air-conditioned buildings, and both engines have been validated by BESTTEST (Delsante, 1995a, b, 2004). Moreover, both softwares have been developed for HERS, which is the main concern of this book.

4.1.2.3 Climate

The interaction between the outdoor climate and the indoor environment is a major concern in the context of the thermal performance of buildings. Climate has a major impact both on the energy and environmental performance of buildings, and also on the comfort sensation of a building's occupants. Obviously, then, the characteristics and also the impact of climate on the occupants and building behaviour depend on the climatic parameters. The main climatic parameters which have to be taken into account when designing a building are: air temperature, humidity, wind, solar radiation, and microclimate (Givoni, 1976; Markus and Morris, 1980; Olgyay, 1963).

Currently, a simple basic classification of climate is used in many nations to apply to building design. It is based on the nature of human thermal requirements in each particular location (Szokolay, 2004). The four main classifications of climates are: cold, temperate, hot-dry and warm humid. The major problem in a cold climate is the lack of heat or excessive heat dissipation for most of the year. Temperate climates have seasonal variations between under- and overheating, but these are not severe. Overheating is the main problem in a hot-dry climate, which has large diurnal temperature variations. The diurnal fluctuation of temperatures is less in hot humid climates. Overheating also is not as great as in hot–dry regions, but high humidity aggravates the sensation of temperature because of restricting the evaporation potential. From this classification it is clear that a moderate climate is most likely to have potential for the design of free running houses.

A moderate climate in Australia has therefore been chosen to evaluate and compare the thermal performances of typical houses in the two different operation modes. A description of the Australian climate follows.

Australian Climate

This continent covers a wide range of latitudes from 9 to 43°S and so it encompasses a significant range of climates. Only the "very cold climate" is not present in the broad climate classifications. The alpine region of Australia has a climate close to this classification, but its population is very small.

Three major climate types, namely hot–humid, hot–arid and temperate (Drysdale, 1975), have been identified for the purpose of building design in Australia. This classification, which was originally based on max/min temperature, humidity and precipitation, was later expanded into several sub-zones. Figure 4.8 shows the distribution of six regions on the basis of temperature and humidity in Australia.

The characteristics of the temperate zone in Australia have been described as follows by Drysdale (1975):

Summer: High daytime, dry bulb temperature (30–35°C)
　　　　　Moderate dry bulb temperature at night (13–18°C)
　　　　　Moderate humidity (30–40%)
Winter: Cool to cold days (10–15°C)
　　　　　Cold nights (2–7°C)
General: Rainfall throughout the year, with winter maximum, except in northern N.S.W. Considerable diurnal temperature range (11–16°C) and a seasonal range of about 16°C.

Both under-heating and overheating, depending on the season, can be a problem but neither is severe.

Climatic Zones and Data for Modelling Samples

As described in Chap. 2, a moderate climate may be described as a critical climate, as energy requirements cannot be accurately predicted, with uncertainty in this regard being greater than in severe climates. Moreover, moderate climate zones are potentially more amenable to the design and employment of free running houses. Therefore a House Free running Rating Scheme (HFRS) is likely to be most appropriate for this type of climate in promoting efficient architectural design and reducing energy requirements, particularly as the population in this zone is greater than in other regions. This climatic condition has therefore been chosen as the context of analysis for the development of a HFRS.

For the purpose of evaluating the thermal performance of buildings, hourly climatic data are required. The Nationwide House Energy Rating Scheme defines 28

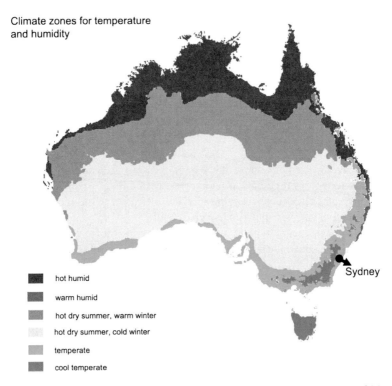

Fig. 4.8 A broad classification of Australian climates based on Australian Bureau of Meteorology data

climatic zones based on hourly data throughout Australia, which are used for simulations in computer programs. Although this climatic classification seems to be enough for the purpose of building efficiency design, the Australian Greenhouse Office has reviewed the weather data from more local weather stations in order to expand the selection of climatic zones and improve the statistical correlation with average data (Lee and Snow, 2006). Recently climate zones have been divided into 69 categories (Lee and Snow, 2006), which can be employed for simulation in any relevant software. The parameters which are included in the hourly data file are: dry bulb temperature (°C), absolute humidity (g/kg), wind speed (m/s), cloud cover (oktas), diffuse irradiance on horizontal (W/m^2) and direct irradiance normal to sun (W/m^2).

4.1.2.4 Thermal Comfort Conditions

As described under Sect. 4.3.1, a thermal comfort condition refers to a condition in which 80–90% of occupants feel thermally comfortable. This condition is affected by both environmental and personal variables. The boundaries of this condition are not the same for free running as for conditioned buildings. In the following the boundaries of thermal comfort are defined for both free running and conditioned houses in the moderate climate of Sydney.

Comfort Condition Boundaries for Free Running Residential Buildings

Adaptive comfort models are arguably more relevant for determining thermal neutrality in houses in the free running operation mode. No significant difference (approximately 1°C) was observed, as shown in Fig. 3.2, between the ranges of temperature and the range used for thermal neutrality proposed by different scholars; thus, for the purpose of a *comparative study* on the thermal performance of houses, all proposed equations are applicable.

The boundaries of thermal neutrality in this study have been defined on the basis of ASHRAE 55(2004), as proposed by de Dear (ACS) (Fig. 3.3). The thermal neutrality limits have been set as 90% of occupant acceptability, for which the range of the comfort zone has been set at 2.5°C on either side of the optimum comfort temperature. It has been defined separately for each month, based on hourly climatic data (Fig. 4.9). These ranges have been taken as the comfort conditions in the living zone of houses occupied during the day-time. The limits for the bedroom zone, during sleeping time (between 12 and 6 am), have been defined as being 5°C less than these bands. The logic is that occupants can easily use a blanket during this time without any complaints about indoor temperature. Other studies have also applied a wider range of thermal neutrality for sleeping areas. For example, the AHRC project

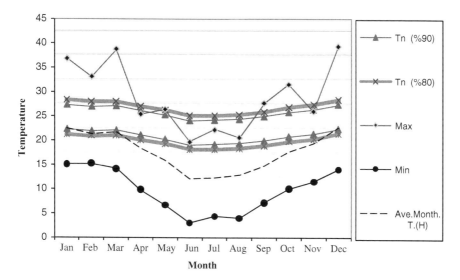

Fig. 4.9 DBT, max and min temperature for each month, and thermal neutrality comfort band for the Sydney climate

Table 4.3 Average monthly temperatures based on hourly temperature and max min monthly temperature, and thermal neutrality for 90% occupant acceptability in the Sydney climate

Sydney Month	Based on hourly data			Based on max and min data				
	Average T (H)	Tn (H)	Tn (90%)	Max	Min	Ave. Tem	Tn (optimum)	Tn (90%)
Jan	22.4	24.7	22.6–27.2	36.8	15.1	25.9	25.8	23.3–28.3
Feb	21.2	24.3	21.8–26.8	33.1	15.2	24.1	25.2	22.7–27.7
Mar	21.6	24.5	22.0–27.0	38.7	14.1	26.4	25.9	23.4–28.4
Apr	18.5	23.5	21.0–26.0	25.4	9.8	17.6	23.2	20.7–25.7
May	15.8	22.7	20.2–25.2	26.4	6.7	16.5	22.9	20.4–25.4
Jun	12.0	21.5	19.0–24.0	19.7	3	11.3	21.3	18.8–23.8
Jul	12.3	21.6	19.1–24.1	22.2	4.4	13.3	21.9	19.4–24.4
Aug	12.9	21.8	19.3–24.3	20.6	4	12.3	21.6	19.1–24.1
Sep	14.8	22.3	19.8–24.8	27.7	7.2	17.45	23.2	20.7–25.7
Oct	17.6	23.2	20.7–25.7	31.5	10	20.75	24.2	21.7–26.7
Nov	19.3	23.8	21.3–26.3	26	11.5	18.75	23.6	21.1–26.1
Dec	22.6	24.8	22.3–27.3	39.3	14	26.0	28.5	29.5–23.5

(Walsh and Gurr, 1982) calculated a lower band of comfort temperature for the sleeping zone, at 5°C less than its range for the living zone.

Table 4.3 shows the average monthly climate data for Sydney. Average mean monthly temperature is calculated on the basis of hourly and max/min monthly temperatures.[3]

The range of optimum thermal neutrality over a typical year was observed to be: 20.7°C ≤ Tn ≤ 25.7°C

Environmental temperature is used rather than air temperature in evaluating the thermal performance of houses. During the summer months environmental temperature computed by AccuRate software is 1 K higher than dry bulb temperature, but during the cold months the two temperatures are approximately the same. Environmental temperature is more reliable than dry bulb temperature in the evaluation of thermal comfort conditions, because the actual thermal sensation of occupants is affected by environmental temperature.

The Effect of Humidity and Airspeed on the Sensation of Indoor Temperature

Although the adaptive model proposed in ASHRAE 55-2004 does not require the inclusion of humidity and air speed, one cannot ignore the effects of humidity on temperature sensation in a humid climate. While humidity is not a major factor in a moderate climate, it cannot be ignored in the Sydney climate in which humidity reaches 80% at times (Australian Bureau of Meteorology, 2006).

The limit boundaries of thermal neutrality have been changed in this study on an hourly basis in response to relative humidity. Certain methods have been proposed to

---

[3] For more accuracy in calculating thermal neutrality and the thermostat setting, the mean monthly temperature should be computed for hourly temperatures rather than max/min temperatures.

4.1 Building Performance Evaluation

include the effect of *humidity* on the limits of comfort temperatures in the sensation of air temperature (Auliciems and Szokolay, 1997; Sutherland, 1971). The effect of humidity has been taken into account in accordance with the ASHRAE standard effective temperature line, and by employing the following (Eq. 4.1) simplified equation proposed by Szokolay (1991).

$$T_{intercept} = T + 23*(T - 14)\ 8\ HR_T (°C) \qquad (4.1)$$

where HR_T is the humidity ratio at temperature T and 50% RH.

Indoor humidity has been considered approximately similar to outdoor humidity in free running houses. However, in reality indoor relative humidity can be lower or higher than outside humidity (Hyde, 1996), depending on the climate.

The effect of *natural air ventilation* is accounted for by the AccuRate software. In summary, the comfort band for free running houses was calculated as follows:

- Thermal neutrality plus or minus 2.5°C for the living area
- The lower band limit adjusted downward for sleeping time by 5°C
- The effect of humidity included to compensate for its effect on the sensation of temperature

Thermostat Settings for Buildings in Conditioned Operation Mode

The notion of thermal comfort in conditioned buildings is implied in the thermostat settings. The thermostat settings indicate when heating and cooling is switched on in the computer simulations. This plays an important role in predicting energy requirements for space heating and cooling. There are several methods to determine the thermostat settings for conditioned houses (Williamson and Riordan, 1997). Different strategies for discretionary heating and cooling of houses result in different predictions of energy requirements, and this issue is particularly critical in temperate climates.

Thermal neutrality as a reference for thermostat setting in AccuRate software is calculated on the basis of mean January temperature, rounded to the nearest 0.5°. The thermostat setting for cooling is 24.5°C and for heating living space is set to 20°C everywhere in Australia. For bedrooms, the heating thermostat setting is 18°C, and over sleeping time (0–7 am) it is 15°C.

The settings were left unchanged for simulating the thermal performance of typical houses in conditioned operation mode, in order to maintain a controlled study, and to enable a comparison between the evaluations of house thermal performances on the basis of the current HERS and of the HFRS to be proposed.

Heating and Cooling Condition

For this study, all conditioned zones were taken to be heated and cooled to maintain the indoor temperature in the comfort band over the occupied time. The occupied time was taken to be the same as that for the free running mode. Service areas, such as laundry, bathroom, store room and garage were not heated and cooled at all.

Heating and cooling are invoked in AccuRate[4] when they are required. Heating is applied for a conditioned zone if its environmental temperature at the end of an hour without heating is below the heating thermostat setting. Cooling is applied if the zone at the end of an hour without cooling or ventilation is outside (above or to the right side of) the bounds of thermal comfort. The boundaries of the comfort region are determined as being between 12 g/kg absolute humidity (AH) at the top of the range, 0 g/kg AH at the bottom and the ET* line based on cooling thermostat +2.5° at the right. If the zone temperature is above the outdoor temperature, ventilation is turned on, and a new temperature and air speed are calculated. If the air speed is above 0.2 m/s, the described comfort region is extended in two ways: the 90% relative humidity (RH) line is considered for the top boundary, and the right boundary is an ET* where:

$$T = 6(V - 0.2) - 1.6(V - 0.2)^2 \quad V \text{ is estimated indoor speed (m/s)} \quad (4.2)$$

If the conditioned zone is still outside the comfort bounds, the zone openings are closed and cooling is invoked, and so the zone temperature at the end of the hour is the same as the cooling thermostat setting.

The samples were considered to be air conditioned when they were simulated in the conditioned operation mode, so during cooling dehumidification was invoked as well. The total annual energy consumption reported is therefore the total predicted energy requirement for cooling, heating and dehumidification.

4.1.2.5 Occupancy Scenarios

As discussed in Sect. 2.4.4, a house performance depends on how its occupants run the building. Occupant behaviour is not predictable; thus most house/home rating systems employ a typical one-occupancy scenario for evaluating thermal performance. However, establishing multiple occupancy scenarios for such a rating system can increase its accuracy.

For this purpose, occupancy scenarios should be defined as parameters that *affect the thermal performance of houses*, rather than how occupants evaluate the particular performance of a house. For example, the type of clothing of occupants should not be an important issue in this definition, since it involves personal evaluation. But the period of time when a house is occupied is a key parameter because over this time it is important for the thermal performance of a house to be acceptable for its occupants, since that is when occupants may turn on the air conditioner. The following parameters therefore are the simplest key parameters for establishing occupancy scenarios.

- Occupation time
- Occupied zones

---

[4]This information is based on the AccuRate manual and the help option in the software (2005).

4.1 Building Performance Evaluation

Occupancy scenarios could be established on the basis of these two key parameters from:
- Surveys or
- Probabilistic scenario based analysis

Surveys are an appropriate way to establish the time when a house may be occupied by different family types. Since the occupation time depends on the family type, statistical information of households and family types of any nation can be taken from its Bureau of Statistics. For example, information taken from the Australian Bureau Statistics (ABS)[5] shows the number and demographic characteristics of people living in families, the relationships between family members, and the types of houses that families live in. It describes the usual resident population of Australia and how the Australian people use their time. However, this statistical information was not sufficient to establish a correlation between family types and house occupation times.

A study by Foster (2006) using ABS information proposed an "occupants factor" to deal with the question of occupancy scenarios in response to criticism of the AccuRate software. By using a section from the ABS survey entitled "How Australians use their time",[6] he proposed multiple weightings to occupancy profiles by day of week. The weightings were determined with regard to the percentage of time that a house might be occupied over a weekday and at weekends, and would be applied in aggregating the results of multiple simulations based on these profiles.

Nevertheless the proposed method, which is conceived as being based on a survey, does not reflect occupancy appropriately, because it does not include differences between occupied zones of a house, which is an important parameter affecting the evaluation of thermal performance. Moreover, the proposed weightings cannot be accepted as standard, since the numerical value of these factors may vary in different countries and in the future, as family types are changing in modern society.

A probabilistic method seems to be a possible solution for determining multiple occupancy scenarios. Probabilistic occupancy scenarios could be established on the basis of the above two key parameters (time and zone). There could be at least $24! \times 3!$ scenarios, in which the first number refers to 24 h per day and the second number refers the minimum number of conditioned zones in a house, namely the bed zone, living zone and one other conditioned zone.[7] However a more manageable number of scenarios would have to be established for application in a future building evaluation system.

In order to come up with a manageable number of synthesized probabilistic scenarios, it is necessary to divide a typical day and occupied conditioned rooms into certain categories. A typical day can be divided into the four categories of wakeup,

---

[5] (ABS) http://www.abs.gov.au/

[6] Australia Bureau Statistic (ABS) "Time Use Survey" Cat. No. 4153.

[7] Note: The three conditioned zones are the minimum zones in a house in which the occupied time and occupants' activities differ. This is the condition considered in some HERS such as NatHERS.

Table 4.4 Multiple occupancy scenarios in this study

Zone	Living zone				Bed zone			
Scenarios	0–6	6–12	12–18	18–24	0–6	6–12	12–18	18–24
Scenario 1		*	*	*	*			
Scenario 2		*	*	*	*			*
Scenario 3				*	*			
Scenario 4		*	*	*	*	*	*	*
Scenario 5		*		*	*	*		*
Scenario 6			*		*		*	

daytime, evening and night time. Therefore, the 24 h of a day are grouped into four categories of 6 h each. With regard to the typical activity of occupants at home (living and sleeping), all conditioned spaces in a residential house are classed into two conditioned zones: the living and the bed zone. With four groups of time and two main conditioned zones, $48 = (4! \times 2!)$ scenarios could be developed for HRS. Each scenario will show the probability of each zone being occupied at a particular time.

A few scenarios were selected in this study to examine how their different effects might be significant in evaluating and rating house performances, particularly in evaluating a lightweight house against a heavyweight one. For this purpose, six scenarios of the 48 probabilistic occupancy scenarios were employed, as shown in Table 4.4.

These six scenarios were selected on the basis of a small investigation of the family and householder types reported by the ABS. The first scenario involves families who use bedrooms only for sleeping. The second is similar to the standard scenario defined in the current rating scheme in Australia, namely NatHERS. The third scenario relates to families who are a couple, and are not at home during the day; however, it does not include the weekend and holiday occupation times. Through scenario four, the performance of a building is evaluated for full-time occupation. Scenarios five and six were considered specifically to investigate the performance of heavyweight and lightweight buildings over the times when lightweight buildings are likely to show better performance.

The importance of the above occupancy scenarios in HRS is described in (Kordjamshidi et al., 2005b, 2009). These two studies show the importance of the occupied time and occupied zone in determining occupancy scenarios, and demonstrate how different occupancy scenarios may affect the result of ranking in a house performance evaluation system.

4.1.2.6 Indicators

As explained under Sect. 3.3, energy ($MJ/m^2.annum$) requirements and weighted exceedance hours called Degree Discomfort Hours (*DDH*) were determined as indicators to evaluate the thermal performance of houses in the conditioned and free running operation modes respectively. Energy requirements were predicted through

4.1 Building Performance Evaluation

simulation by AccuRate software with a house run in the conditioned mode. To calculate DDH, an algorithm was designed in order to convert hourly temperature data produced by AccuRate software into DDH.

Post processing was carried out to calculate the DDH. The process of calculating DDH is shown in Fig. 4.10. The simplified algorithm was written in Excel to convert

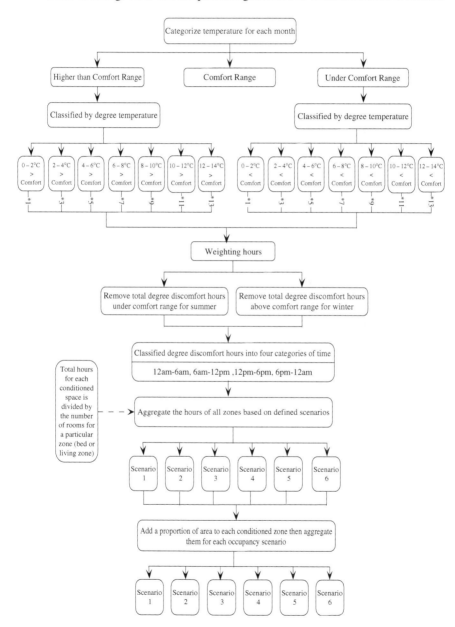

Fig. 4.10 Process of computing DDH

computed annual indoor temperatures (8,760 h) for all zones of a house into the respective DDH indicators.

As the typical houses differed in the number of their rooms and area (m^2), 6 typical files were required to be designed for the 6 base samples in order to compute DDH for each simulated model. Each Excel file contained a number of separate sheets matched with the number of conditioned zones. Hourly temperatures for each simulated house in the free running mode were imported into the related Excel file in order to calculate its annual DDHs. The outcome contained two indicators of DDH; one being DDH with area weighting, which is weighted by proportion of area, and the other being DDH without area weighting.

The purpose of considering an area weighted DDH was to better investigate the correlation between two identified indicators of house performances: MJ/m^2 and DDH with weighted area, in two different house operation modes. The DDH of each living area and bed zone (conditioned zones) was given a weighting in proportion to its area as a fraction of the total area. The sum of the area weighted DDHs of conditioned zones provided the "DDH with area weighting".

However, it was assumed that degree discomfort hours would be free of the effect of floor area, so as to avoid the same probabilistic problem which was observed in the energy based rating schemes discussed in Sect. 2.4.3.

4.1.3 Effective Parameters for Improving Buildings Thermal Performance

Building design features are the most important parameters affecting thermal performance of buildings. This fact is recognized and demonstrated by many researchers (Hyde, 2000; Planning, 2006; Tavares and Martins, 2007; Tuhus and Krarti, 2010; Willrath, 1997; Zhai and Chen, 2006). Some of these parameters are described in the following:

- Internal wall

  Internal walls are a partition from which convective heat flows between zones and so they affect the indoor temperature of a conditioned zone. An interior wall between two zones with a large temperature differences behaves like an exterior wall (Akbari et al., 1986). Thus the mass and construction of indoor walls can significantly affect thermal sensation and also heating and cooling energy requirements.

- Insulation

  Heat conduction through building envelopes noticeably affects the fluctuation of indoor temperature and is a major component of the cooling and heating energy load, so that the indoor climate effectively depends on the resistance of a building's external surface.

- External Colour

  The colour of the outside surface of a building envelope influences the thermal performance of a building because it determines the amount of absorbed

4.1 Building Performance Evaluation

solar radiation and its inward transmission into a building. Dark external colours improve performance in the winter as they increase the absorbance of solar energy, but reduce summer performance.

- Building orientation

  Building orientation can also affect the indoor climate of a zone or room because of its effect on the ventilation and solar absorbance of external walls. Orientation is relatively more important with respect to wind than to the patterns of solar irradiation, as discussed by Givoni (1976). Since natural ventilation is a key parameter in improving the performance of buildings, particularly in free running operation, the magnitude of orientation as a factor cannot be ignored for an efficient architectural design.

- Infiltration

  Air infiltration through cracks results in heat gain and heat loss; therefore infiltration or an uncontrolled ventilation rate reduces the thermal performance of a building.

- Glazing type

  The effect of the glazing type used in the transparent parts of a building on the performance of the building has been well documented (Klainsek, 1991; Nielsen et al., 2001; Omar and Al-Ragom, 2002), particularly for air conditioned buildings. Glazing systems have a huge impact on the performance of a building, and glazing modification often presents an opportunity for indoor climate improvements in a building. Window glazing and frames are a major factor in determining the energy efficiency of the building envelope.

  The three main types of glass are clear, reflective and tinted. Reflective windows diminish solar transmission without heating up the glass area. A large proportion of the radiation can be absorbed by tinted glass when the glass surface is heated up. The heat is then radiated and conducted to the surrounding area, so for summer the benefit of this type of glass, compared to reflective glass, is reduced.

  Single glazed windows provide only a small amount of insulation to the passage of heat because of thin films of still air that exist next to the glass. Double glazing gives better insulation because there are two panes of glass with a sealed space between them.

- Internal window covering

  Window coverings present a different effect on the performance of a building depending on their placement, whether internal or external. An internal window covering limits the glare resulting from solar radiation. This device, depending on its material and colour, can be effective in minimizing uncomfortable glare from direct beams, and in reducing the sun's heat from entering the space. Moreover, it increases the heat resistance of glazed areas when closed.

- Sun shading devices

  External shading devices affect the thermal performance of a house by reducing the incident solar intensity and therefore incident solar energy, or the amount of solar energy on windows and walls. Shading can be created by overhangs or eaves and vertical side fins, but the former have been shown to be more effective in shading than the latter (Offiong and Ukpoho, 2004). However their effectiveness depends on their width.

- Ratio of openable windows

  The presence of openable doors and windows affects air ventilation and improves the thermal performance of a naturally ventilated building when the outdoor temperature is less than the indoor temperature during summer. Depending on the climate, increasing its ratio to that of overall glazing can also significantly reduce the cooling load.

- Window to wall ratio

  Windows have been described as "thermal holes" (Fisette, 2003). They are mostly 10 times less energy efficient than the wall area they replace. Thus, depending on the window area and its orientation, a house can lose 30% of its air-conditioning energy. However their effect on the indoor environment of a naturally ventilated building will differ from that on a conditioned building.

These parameters were identified as being of primary significance in investigating both the energy requirements and DDH in typical houses in the selected climate zone. Thus the effect of each of these parameters was tested, to investigate whether their individual effect was similar in the performance of a house in conditioned and in free running operation mode. The variation of each parameter employed in the simulated modes is shown in Table 4.5

Table 4.5 House parameters for simulations

	Parameter descriptions	Variation of parameter
1	Ceiling insulation (resistance)	0, 1, 2, 3, 4 (m^2 K/W)
2	Wall insulation (resistance)	0, 1, 1.5, 2, 3 (m^2 K/W)
3	Floor insulation (resistance)	0, 1, 1.5, 2 (m^2 K/W)
4	Internal wall construction	Plasterboard, concrete block, brick plasterboard and cavity brick
5	Infiltration (air change per hour)	0, 1, 2, 5
6	Window covering (resistance)	Open weave (0), closed weave (0.03), heavy drape (0.055) and heavy drape + pelmet (0.33)
7	Percentage of open able window (%)	25(base), 50%, 75%
8	Shading device (eave length)	0, 450, 600, 1,000 mm
9	Orientation (degree)	0, 45, 90, 135, 180, 225, 270, 315
10	Glazing type (shading coefficient)	Single glazing (1), reflective (0.52), tone and clear (0.70), Double glazing (0.88), clear and tone (0.60)
11	Roof colour (absorbance)	Light (30%), Medium (50%), Dark (85%)
12	Wall colour (absorbance)	Light (30%), Medium (50%), Dark (85%)
13	Window to wall ratio (N & S) (%)	0(base), 15%, 25%
14	Window to wall ratio (E & W) (%)	0(base), 15%, 25%

0 (base) refers to the percentage of window to wall ratio in the typical houses which is different for different models. It does not mean that the ratio of window to wall is 0.

4.2 Parametric Sensitivity Analysis of Thermal Performances of Buildings: A Comparative Analysis

4.2.1 What Is Sensitivity Analysis?

Sensitivity is a measure of the effect of change in one factor on another factor. Sensitivity analysis is potentially useful in all phases of the modelling process: model formulation, model calibration and model verification. Tarantola and Saltelli (2003) demonstrated that sensitivity analysis can produce useful information regarding the behaviour of the underlying simulated system. Sensitivity analysis is used to assess the relationship between variations in input parameters to variations in output parameters, and has been used in many studies on buildings (Lam and Hui, 1996; Lomas and Eppel, 1992; Zhai and Chen, 2006), including for assessing their thermal performance and their energy load characteristics.

The aim of sensitivity analysis is to observe the system's response following modification in a given design parameter (Cammarata et al., 1993). However there are no formal rules or well-defined procedures for performing sensitivity analysis for building design, because the objectives of each study will be different and building descriptions are quite complicated. In most cases, perturbation techniques and sensitivity methods are used to study the impact of input parameters on different simulation outputs as compared to a base case situation. The results are then interpreted and generalized so as to predict the likely responses of the system. The concept is simple and straightforward, but a clear understanding of what sensitivity analysis can do for studies of building thermal performance and energy consumption and how the results should be interpreted is very important.

In this case the aim of sensitivity analysis has been to investigate the effects of house envelope variations on the annual and seasonal behaviour of the typical houses studied, and in particular to investigate possible linkages between the thermal performances of a house in different operation modes in response to the variations described above in Sect. 4.1.3.

The outcomes of the comparative simulations are explained in the following four categories for each parameter:

- house operation mode (free running/conditioned)
- house construction (heavyweight (HW)/lightweight (LW))
- house type (single storey (SS)/double storey (DS))
- seasonal performances (summer/winter)

4.2.2 Thermal Performances of Dwellings in the Sydney Climate

The following section reports the result of the simulations for the Sydney climate and their analysis. Observations are reported based on the absolute value or relative (percentage) changes in the predicted thermal performance of base cases resulting from a modification of each one of their design parameters (presented in Table 4.5). As the main concern of the comparisons is to find any difference between the thermal performances of houses in each of these categories, differences are noted even if the numerical value of changes in the thermal performance may not be significant.

Note: As previously defined in Sect. 4.1.2.6, thermal performance of free running houses was evaluated on the basis of Degree Discomfort Hours (DDH), with and without area weighting. In a small study by the author on the thermal performances of the typical houses in free running operation mode, no significant difference was observed between the evaluation of buildings on the basis of DDH with and without area weighting. Therefore, calculating DDH without area weighting is more likely to be appropriate for the purpose of rating, to avoid any discrimination between bigger and smaller houses. Consequently the result of an evaluation of the thermal performance of free running houses in terms of DDH with area weighting is not shown here.

4.2.2.1 Ceiling Insulation

The effect of three different bulks of ceiling insulation on the thermal performance of the typical houses, with both lightweight and heavyweight construction, was simulated. The results were as follows.

Free running and conditioned mode. In both free running and conditioned mode the thermal performance was found to be improved by the use of bulk ceiling insulation as shown in Fig. 4.11. The addition of R2 insulation to the ceiling of a typical house gave an average reduction in annual energy requirements of 42.5%. In free running mode, the annual degree discomfort hours was decreased by 32.9%. The application of insulation greater than R2 produced a minimal change in the thermal performance of both house modes.

Employing ceiling insulation as a technique for improving design efficiency, therefore, creates a greater reduction in annual energy requirements in conditioned houses than in the annual discomfort degree hours in free running houses.

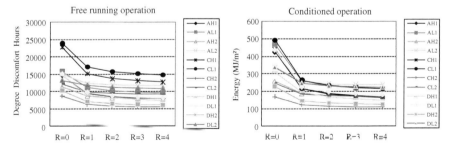

Fig. 4.11 Projected effect of ceiling insulation on the annual thermal performances of typical houses in conditioned and free running operation modes

4.2 Parametric Sensitivity Analysis of Thermal Performances of Buildings

Seasonal performance. Ceiling insulation improved both summer and winter performance in both operation modes. This is illustrated in Figs. 6.1 and 6.2, Appendix. A greater improvement in the seasonal performance of houses in conditioned mode was found in both summer and winter performances when R2.0 insulation was added to the ceiling, with an average improvement of 41–44% respectively. In free running mode the improvements were 32 and 42% respectively. This indicates that the seasonal performance of houses in the conditioned mode is more sensitive to any change in ceiling insulation.

Single Storey and Double Storey. Although adding ceiling insulation affected the thermal performance of both SS and DS houses, this effect was greater in SS houses, whether the house was in free running or conditioned mode. The application of R3 insulation improved the annual thermal performance of SS houses approximately 1.5 times more than it did in DS houses in both modes. The main reason for this appears to be the typical placement of the bed zone above the living zone in the DS houses. This particular arrangement substantially isolates the living zone ceiling from the outdoor environment. Accordingly, heat transfer between the living zone and the outdoor environment through the ceiling is reduced and ceiling insulation does not change this phenomenon. Since the thermal performance of the living zone (in DS) is not greatly sensitive to ceiling insulation, and the sensitivity of a house's thermal performance depends to a great extent on the sensitivity of its living zone, the thermal performance of the DS houses does not greatly change when an insulation layer is added to the ceilings.

This result may differ in other occupancy scenarios, depending on the number of hours that the living zone is occupied. It could also change if the location of the living zone was transferred from the ground floor to the first floor.

4.2.2.2 Wall Insulation

Wall insulation has been nominated as a useful technique for improving the thermal performance of all construction types in conditioned mode in the Sydney climate (Willrath, 1997). However, this study has found that while adding insulation would enhance the performance of a conditioned house, this was not the case for some houses in free running mode.

Free running and conditioned mode. Fig. 4.12 depicts a situation in which an increase in wall insulation would result in:

- a decrease on annual energy requirements,
- a general decrease on annual degree discomfort hours,
- an increase on annual degree discomfort hours for single storey heavyweight houses.

The addition of R2 insulation to the external walls of typical houses resulted in an average 14% improvement in their annual thermal performance in conditioned mode, but the improvement was only an average of 6.9% for all houses in free running mode. This average improvement does not include Single Storey houses

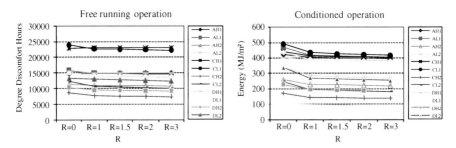

Fig. 4.12 Projected effect of wall insulation on the annual thermal performances of typical houses in conditioned and free running operation modes

with heavyweight construction, as this group of simulated models appeared to be an exception to this general observation.

Heavyweight, single storey houses presented a slight deterioration in their free running annual thermal performances (0.8%) when insulation was added to their external wall. The reason for this overall deterioration is apparent when their seasonal performances are examined, and is discussed in more detail in the relevant section below.

Note: the following comparisons between SS & DS and HW & LW do not include the thermal behaviors of SS HW houses.

Single Storey and Double Storey. DS houses achieved a greater improvement in their thermal performance when insulation was added to their external walls than did SS houses. In free running mode the improvement was an average of 7% more in DS houses; and in conditioned mode it was about 10%. This is shown in Fig. 4.13.

One reason for the greater improvement in DS houses seems to be the placement of the bed zone above the living zone, which performs as an insulation layer above the living zone (above its ceiling) in that typical design of DS houses. With external wall insulation, all external sides of a living zone in a DS house appear to be

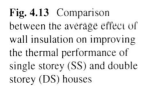

Fig. 4.13 Comparison between the average effect of wall insulation on improving the thermal performance of single storey (SS) and double storey (DS) houses

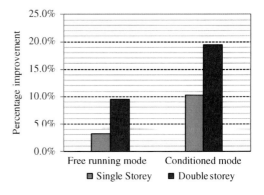

4.2 Parametric Sensitivity Analysis of Thermal Performances of Buildings

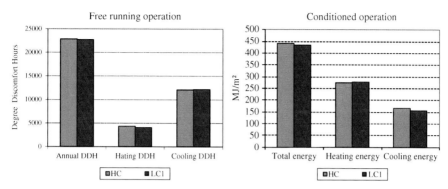

Fig. 4.14 Comparison of the thermal performance of a HW SS house in the Sydney climate without an insulation (HC) layer and a LW house with R.1 insulation in its external walls (LC)

insulated, while in SS house there is more heat transfer through the ceiling. Although the situation of the bed zone in both house types is similar, the thermal performance of a living zone is more important in evaluating annual thermal performance of a house based on the first occupancy scenario, in which 75% of the time the living zone is occupied.

Heavyweight and Lightweight. The typical houses with lightweight construction achieved greater improvement in their thermal performances than those with heavyweight construction in response to the addition of insulation to the external walls. Greater improvement was observed in both conditioned and free running modes. With the addition of R.2 insulation, typical LW houses presented an average of 8.7% reduction in annual degree discomfort hours and 17.5% reduction in annual energy requirements. HW houses showed an average of 4.6% reduction in annual DDH, and 11.3% reduction in energy requirements.

When R1.0 insulation was added to the external walls of LW single storey houses, they achieved an aggregate thermal performance equivalent to that of HW SS ones without insulation (Fig. 4.14). This finding therefore seems to confirm the assumption described in Chap. 2, that LW SS buildings are able to achieve annual thermal performance comparable to those with HW construction in the Sydney climate.

Seasonal performance. The effect of wall insulation on the seasonal performance of typical houses is illustrated in Figs. 6.3 and 6.4, Appendix.

As previously noted, for typical houses in conditioned mode, adding insulation to external walls resulted in a predicted reduction of heating and cooling energy requirements. However, the summer performance of single storey houses with heavyweight construction appeared to be unresponsive to wall insulation. This observation implies that wall insulation would not be a useful technique to improve the thermal performance of single storey detached dwellings in warmer climates.

Also as noted previously, external wall insulation improved the free running winter performance of typical houses, with the exception of single storey, heavyweight houses. The SS, HW winter performances were found to be unresponsive to wall insulation (an average of 0.2% change), while the summer performances

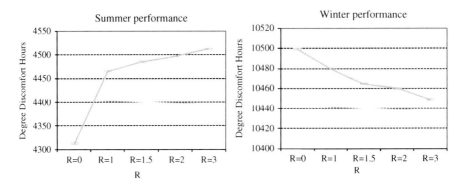

Fig. 4.15 Projected effect of wall insulation on the free running seasonal performance of a SS HW house

deteriorated by about 2%. Thus the slight deterioration that was observed in their annual performance in response to increased insulation would appear to be the result of degradation in their summer performances. This situation is depicted in Fig. 4.15 for a typical SS HW house.

4.2.2.3 Floor Insulation

A slab on the ground floor without carpet gives the best performance in the Sydney climate (Willrath, 1997). However in the housing market most of the floor area is usually covered by carpet. Therefore, in order to simulate typical houses, slab floors in heavyweight houses were simulated as covered by carpet. For HW houses an insulation layer was added under the slab floor.

The timber floor of typical houses with LW construction was suspended 60 cm above the ground. This type of floor was not covered by carpet, and insulation for these cases was added under the suspended floor.

Free running and conditioned mode. In free running mode adding a level of insulation under the ground floor of the typical houses caused a slight degradation in the annual thermal performances of all of them, whereas in conditioned mode, depending on the type and construction of houses, this addition produced varying changes.

As shown in Fig. 4.16, adding R1 insulation resulted on average in 3.9% deterioration in the annual free running performance of the typical houses. In the conditioned mode this addition produced two different patterns of annual energy requirement for SS and DS houses, and will be described in the following section. It is worth noting that in both house modes a higher level of insulation produced a minimal change of less than 1% in their annual thermal performances.

Heavyweight and Lightweight. Overall patterns of thermal performance of the typical HW houses showed no significant sensitivity toward the addition of under-floor R1.0 insulation. The houses with LW construction achieved a slight improvement in the annual thermal performance in the conditioned mode (average

4.2 Parametric Sensitivity Analysis of Thermal Performances of Buildings 83

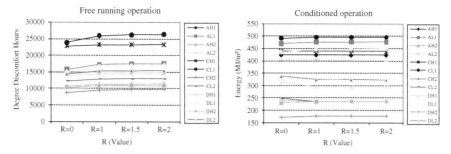

Fig. 4.16 Projected effect of floor insulation on the annual thermal performances of typical houses in conditioned and free running operation modes

1.15%), but an apparent deterioration in their free running performance (average 6.5%). Possible reasons for the latter unexpected observation are explored in the following section.

Single Storey and Double Storey. In the free running operation mode no significant difference was observed between sensitivity in the thermal performance of DS and SS houses in response to the insertion of under-floor insulation. In the conditioned operation mode a marginal deterioration (0.8%) was observed in response to the addition of R.1 insulation under the floor, while the same change made a slightly greater (2.4%) improvement in the performance of DS houses. The different effect of floor insulation appears to be attributable to more improvement in the winter performance of DS houses and will be described in the following.

Seasonal performances. The effectiveness of floor insulation was investigated in the seasonal behaviour of the typical houses. Figures 6.5 and 6.6 in Appendix depict the patterns of seasonal behaviour of the houses in response to the addition of three different levels of under-floor insulation.

The insertion of an R.1 insulation layer under the floor of uninsulated typical houses resulted in an apparent deterioration in both seasonal performances in the free running mode, with an average of 8.3 and 5.3% deterioration in the summer and winter performances respectively. The responses in the winter performance become more interesting when we note that deterioration in the winter performance of LW houses (5.8%) was significantly more than that in HW houses (0.014%). A reason for this phenomenon is the effect of those times when the temperature in the suspended floor zone is higher than the indoor temperature of spaces on the ground floor, such as the living zone. In this situation, insulation reduces the benefits of subfloor temperatures by reducing heat transfer between subfloor and ground floor. The likelihood of this explanation being correct was tentatively confirmed by examining the zone temperatures from the simulation outputs for a relevant day.[8] This

---

[8] A comparison was made between the hourly temperature of the living zone and the suspended floor for one of the LW typical houses in the free running operation mode when R2 insulation was added under the suspended floor. The following figure shows the hourly zone temperature for a warm and cold day respectively, on the 15th of January and of July. Both figures illustrate

finding points to the inapplicability of floor insulation to improving the thermal performance of a free running house.

The winter performance of a conditioned house can be improved overall by insulating its floor. Applying R.1 insulation under the floors of typical houses in conditioned mode resulted in:

An average of 2.8% reduction in their heating energy requirement. The reduction in DS houses (4.9%) was greater than that in SS houses (0.8%).

An average of 3.5% increase in their cooling energy requirement. This was the same for both SS and DS houses.

Therefore the observed improvement in the annual thermal performance of DS houses occurred because the improvement in the winter performance of these houses outweighed the deterioration in their summer performances.

4.2.2.4 Wall Colour

External surface colour has a significant effect on the thermal performance of buildings. This is shown in a theoretical and experimental study in Bansal et al. (1992). Dark and light colours reduce under-heating and overheating respectively. That is to say, a light external colour is effective in reducing cooling energy requirements in a warm climate. In a Sydney study, when dark external colours were replaced by light colours, the annual energy requirement of an uninsulated house with standard mass[9] was reduced by 13% (Willrath, 1997). Though the light external colour enhanced the annual thermal performance of a conditioned house, this colour did not cause an enhancement in the annual thermal performance of the same house in free running mode. This will be explored in the following.

---

that the temperature of the suspended floor zone is closer to thermal comfort temperature than the temperature of the living zone for more hours of the day. Floor insulation restricts heat transfer between these two zones and therefore in effect reduces the benefits of a suspended floor for a free running building in such ambient conditions. Comparison between temperature of living zone and sub floor zone of a lightweight house for 15th July and January in the Sydney climate.

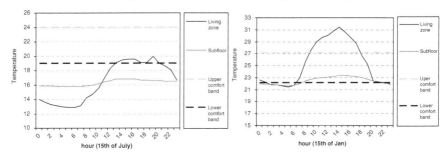

Comparison between temperature of living zone and sub floor zone of a lightweight house for 15th July and January in the Sydney climate.

[9]Standard mass refers to a house with brick veneer RFL in external walls, R2.5 in the ceiling and no floor insulation.

4.2 Parametric Sensitivity Analysis of Thermal Performances of Buildings   85

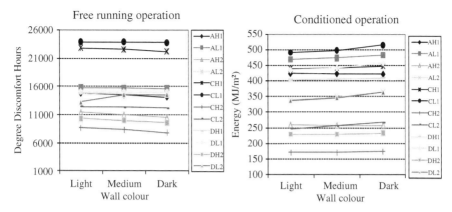

Fig. 4.17 Projected effect of external wall colour on the annual thermal performances of typical houses in conditioned and free running operation modes

Free running and conditioned mode. The colour of the external wall had various effects on the thermal performances of houses in different modes. Figure 4.17 clearly demonstrates these differences. When light colour on the external wall was replaced by dark colour, an average of 3.5% enhancement occurred in the thermal performances of typical houses in the free running mode. The same change produced an average of 3.6% deterioration in the thermal performance of the houses in the conditioned mode. Seasonal behaviour is likely to be the main reason for this variation.

Seasonal performance. Applying dark colour with 85% absorbance instead of light colour on the external wall improved a house's winter performance and degraded its summer performance in both modes. Figures 6.7 and 6.8, Appendix illustrates this situation among the typical houses.

The amount of improvement and apparent deterioration in the seasonal performance of an uninsulated house depended on the operation mode. The summer performance was more sensitive to changing external wall colour in the conditioned mode than in the free running mode, whereas the winter performance in free running mode was more sensitive to such changes. With the application of a dark instead of light external wall colour, the deterioration on the summer performance of houses in conditioned mode was found to be more than that in the free running mode (about 7%). The same change, by contrast, gave slightly more improvement in the winter performance of houses in the free running mode (2%).

The lower degradation in the summer performance of free running houses is related to the capability of such houses to reduce the number of overheating degree hours. The impact of the exterior dark colour is lessened because of the natural ventilation which benefits these houses during hot months. Although this benefit exists in conditioned mode,[10] it is not as significant as it is in the free running mode.

---

[10] AccuRate software counts the effect of air ventilation for prediction of cooling energy requirements (see se Sect. 4.1.2.2 for more clarification).

The different effects of the external wall colour that are seen in the annual thermal performances are due to the high degradation in its summer performance in the conditioned mode. This means that advice appropriate for improving the performance of the conditioned houses may not necessarily be applicable in the case of the free running operation mode.

Heavyweight and Lightweight. The impact of external wall colour on the thermal behavior of a house was related to the construction of the building. HW houses showed more sensitivity to changes in the external colour of the wall in free running mode, with the thermal performance improving by 5.8%, while LW houses displayed more sensitivity to the same change in conditioned mode, with a deterioration of 6% in thermal performance. This observation again reinforces the view that the house operation mode is a significant consideration when making a decision about adopting appropriate advice for efficient architectural design.

Single Store and Double Storey. There was a greater change in the thermal performance of DS houses than in SS houses in response to variation in external wall colour. An average of 6.1% reduction occurred in the heating energy requirement of SS houses with the replacement of a dark colour for a light one, while the reduction was roughly 1.5 times greater (9.2%) in DS houses. This observation was similar whether the houses were in the conditioned or free running operation mode. The reason seems to be due to the typical design of DS houses in this study, as explained previously, for which the effect of any changes on the external wall, such as wall insulation and wall colour, produces more changes in the thermal performance of DS houses.

4.2.2.5 Roof Colour

Absorptance is the thermal property related to material surface and colour. The absorptance of a roof has a considerably greater effect on energy loads than external walls do (Shariah et al., 1998). Givoni (1976) has pointed out that the external surface of the roof is often subject to the largest temperature fluctuations, depending on what type it is, and on its external colour. The simulations in this study similarly demonstrated a considerable impact from roof absorptance on the indoor environment, particularly in single storey houses and in conditioned mode.

Free running and conditioned mode. A change in the external roof colour from light (30% absorptance) to dark colour (85% absorptance) degraded a house's annual thermal performance by between 4.5 and 44% for all typical houses in conditioned mode. In free running mode different effects on the annual thermal performance were found, depending on whether the houses were single or double storey (Fig. 4.18).

Single Storey and Double Storey. Two different effects were observed among the free running houses from a change in roof colour. The application of a dark colour resulted in a deterioration of 7.4–11.7% in SS free running houses. In the performance of DS houses there was a marginal improvement of 3.7%. The reason for this is that in DS houses in this sample, as before, the bed zone is designed to be above the living zone. Because of this, the number of "cooling degree hours"

4.2 Parametric Sensitivity Analysis of Thermal Performances of Buildings

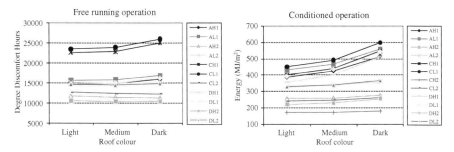

Fig. 4.18 Projected effect of roof colour on the annual thermal performances of typical houses in conditioned and free running operation modes

caused by the effect of solar radiation absorptance on the dark roof is reduced in the living zone. On the other hand, the dark colour advantages DS houses over the cold months, by reducing "heating" degree hours in the bed zone in winter, without affecting overheating of the living zones during the hot months.

Heavyweight and Lightweight. Typical houses with LW construction in free running mode showed more sensitivity to external roof colour change in their thermal performance than houses with HW construction. There was slightly more sensitivity (1–3%) in annual thermal performance of LW houses in this mode than in HW houses when light colour was replaced by dark colour. This situation was reversed in the conditioned mode.

Seasonal performance. Figs. 6.9 and 6.10 in Appendix illustrate a breakdown of the seasonal performance of typical houses in response to the variations in external roof colour. The figures depict an improvement in the winter performances and degradation in the summer performances of houses in both free running and conditioned mode. Free running buildings, taking more advantage of natural ventilation, have the ability to adjust (reduce) overheating caused by a dark roof colour.

In the Sydney climate, therefore, an appropriate external roof colour seems more important for improving the thermal performance of a conditioned house than for a free running house. However it should not be ignored that in free running mode the DDHs may stay high.

4.2.2.6 Orientation

A building's orientation to the sun will impact on the house's ability to optimize passive heating and cooling, and natural ventilation. Solar heat gained through the external surface of a building depends on the orientation of the surface (azimuth). The influence on indoor temperature of orientation, together with its interaction with other building envelope features, has been well documented (Givoni, 1976). Orientation, therefore, is an important parameter in an efficient architectural design. The following investigates any difference between the sensitivity of houses' thermal performances in different operation modes in response to changing orientation.

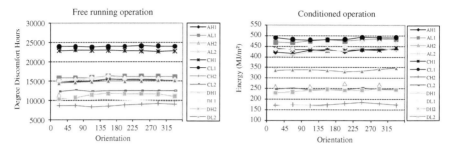

Fig. 4.19 Projected effect of orientation on the annual thermal performances of typical houses in conditioned and free running operation modes

Free running and conditioned. The findings of this study indicate a relatively small variation in the thermal performances of houses in response to changes in its orientation. Figure 4.19 shows the outcomes of changing the orientation by 45° increments (over eight orientations), which produced an average variation of 5% in the annual thermal performance in free running mode, while in conditioned mode the variation was on average 5.5%. The explanation for this small variability is the design of all typical house plans, with relatively equally sized windows, which are distributed on all sides.

The variation depends largely on the changes in the thermal performance of the living zone, because this zone is occupied for the most time (18 h per day) in the first specified occupancy scenario. Therefore, to improve the thermal performance of a house, the highest priority would be to optimize the orientation of the living zone.

Heavyweight and Lightweight. The influence of variations in house orientation on a house with HW construction was found to be greater than its effect on the same house with LW construction, because of the substantial potential of thermal mass to delay transferring heat gains. In response to incrementally rotating the orientation of the typical houses by 45°, the variation in the thermal performances of HW houses was on average 3 times greater than that for houses with LW construction in free running mode, but was only about two times greater in conditioned mode.

Seasonal performance. A breakdown of seasonal performance in response to variations in orientation is given in Figs. 6.11 and 6.12, Appendix. It was observed that the impact of orientation on the variation of "cooling" degree hours and cooling energy (summer performance) was greater than its impact on the variation of "heating" degree hours and heating energy (winter performance). This observation points to the importance of building orientation for taking advantage of wind direction for natural ventilation, particularly for free running buildings, in addition to making use of solar heat gain over winter.

Single Storey and Double Storey. No significant difference was observed between the average changes in the annual thermal performance of DS and SS houses as a result of changing the house orientation. DS houses showed slightly more sensitivity (about 1%) to various house orientations in thermal performance than did SS houses in this study. As mentioned before, the thermal performance of a house depends strongly on the thermal performance of the house's living zone. In DS houses the living zones are generally designed on the ground floor and under the bed zone, and therefore there is no direct heat gain from the roof in this zone. This means that the thermal performance of a DS house is reasonably sensitive to the external wall orientation, whereas the thermal performance of the living zone in the SS houses is significantly affected by the thermal effects of the roof. The fluctuations in the winter performances of DS houses in response to the various orientations depend on the external wall area of the living zone.

4.2.2.7 Overhang Depth

Solar gain through windows is obviously the largest load component, and window shading can have a significant impact on solar loads. Shading provided by overhangs at the top of windows with no offset distance was studied and its effect on energy and degree discomfort hours was calculated.

Free running and conditioned. Adding an overhang above all windows improved annual thermal performance of the typical houses in conditioned mode. However, this was not the case for the houses in free running mode. The houses presented different patterns of annual thermal performance in the different modes, responding to increments in overhang width. This situation is shown in Fig. 4.20. Increases in overhang width of 1 m resulted in an average of 4.8% enhancement in the house's annual thermal performance in conditioned mode. The same overhang generally caused 3.7% deterioration in annual free running performance, while a slight improvement was observed in thermal performance in some cases. The reason for these different results is made clear when comparing the seasonal performances of each house in the two different house operation modes.

Seasonal performance. The setting of overhang at the top of all windows with no offset was beneficial for improving summer performance. The benefit was greater for a conditioned house than for a free running house. When all the overhang widths were increased to 1 m, the summer performance improved by an average of 26.6% in conditioned mode. The improvement was, however, only an average of 8% for all houses in free running mode. The winter performance of the typical houses in this situation deteriorated by an average of 8.2% in conditioned mode and 11.7% in the

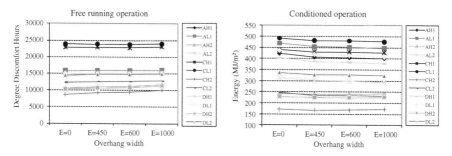

Fig. 4.20 Projected effect of overhang width on the annual thermal performance of typical houses in conditioned and free running operation modes

free running mode. Thus the poorer outcome that was observed in the annual free running performance of some of the houses (in response to the setting of overhang with a width of 1 m for all windows) would appear to owe more to deterioration in their winter performance and less improvement in their summer performance in free running mode than in conditioned mode (See Figures 6.13 and 6.14 in Appendix). Operable overhangs are therefore obviously useful for different house operation modes.

Heavyweight and Lightweight. HW houses were twice as sensitive to changes in the width of overhang as LW houses in free running mode. This result was reversed for LW houses in the conditioned mode.

It should be noted that in an efficient architectural design a suitable overhang width would be designed in relation to window area and orientation. The purpose of the analysis here is to examine the discrepancy between thermal performances in two different building modes in response to the same conditions. Thus a series of overhangs of the same width was applied for all windows, regardless of the orientation.

Single Storey and Double Storey. A comparison was made between the average percentage changes in the annual thermal performance of the DS and SS houses when adding overhang above the windows. It was found that in free running mode the average change in the DS houses was 12 times greater than that in the SS houses. In contrast, in conditioned mode the average annual performance of the SS houses was 1.5 times greater than that in the DS houses. This observation confirms the importance of house types in the interaction with house modes.

4.2.2.8 Glazing Type

The performances of the typical houses were simulated using a range of glazing types. Figure 4.21 shows the patterns resulting from changes in performance when all glazing was changed from single glazing to other specified types (See Sect. 4.1.3 for glazing specifications).

Free running and conditioned modes. Fig. 4.21 depicts the situation of typical houses in annual thermal performance when all their glazing (SG Clr) was

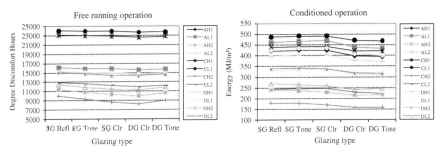

Fig. 4.21 Projected effect of glazing types on the annual thermal performances of typical houses in conditioned and free running operation modes

4.2 Parametric Sensitivity Analysis of Thermal Performances of Buildings

replaced by other specified types. All typical houses in free running mode showed the same patterns in their annual thermal performances in response to the application of different glazing types. These patterns were slightly different from those in the conditioned mode.

The replacement of all SG Clr with SG tone resulted in an average of 4.6% degradation in free running annual thermal performance, while in the conditioned mode there was significantly less degradation (0.3).

DG improved the thermal performance of houses in both modes. In a study by Willrath (1997) replacement of SG with DG reduced the annual energy requirement by between 12 and 17%, depending on the window frame type. The finding in this study showed an average of 6.2% reduction in the annual energy requirement of typical houses, in response to the same glazing replacement, with an aluminium frame. This replacement caused a reduction of 2.1% in annual degree discomfort hours in free running mode.

As noted above, the thermal performance of a house is affected by the type of glazing, whether it is in free running or conditioned mode. Nevertheless, the percentage of degradation or improvement depends on the house operation mode, which is discussed in detail in the following. This phenomenon again highlights the significance of house operation mode in providing particular advice for improving design efficiency.

Double Storey and Single Storey. A comparison of SS and DS performances in free running mode demonstrated that the thermal performance of DS houses was more sensitive to the type of glazing. For instance, by replacing SC Clr with SG Refl. the range of deterioration in the annual thermal performance of DS houses was between 15.5 and 1.8%, while this range for SS houses was only between 2.9 and 0.5%.

The above observation illustrates that the potentiality of DS houses to improve their thermal performance differs from that of SS houses.

Seasonal performance. The influence of four different glazing types on the seasonal performance of the houses is illustrated in Figs. 6.15 and 6.16, Appendix. The illustration shows the patterns of summer and winter performances in both free running and conditioned modes. The pattern is the same for both modes. However, the summer performance of conditioned houses was more affected by variations in glazing. For instance, in conditioned mode the houses achieved an enhancement of 7.67% on average in their summer performance when changing SG Clr to SG tone. This change improved their free running performance by an average of 4.8%. Their winter performance was degraded by 4.5% in conditioned mode and by 3.5% in free running mode.

Heavyweight and Lightweight. The thermal performance of a HW house was relatively more sensitive to changes in glazing type than the thermal performance of the same house with LW construction in both building modes. For instance, replacing all SG Clr glazing by SG tone produced an average of 2.7% deterioration in the annual free running thermal performance of typical houses with HW construction. This deterioration was reduced to 1.1% for houses with LW construction.

4.2.2.9 Window Covering

A building's thermal performance can be improved by using openable internal window covering. This improvement depends on the level of resistance, transmittance and absorptance of the window covering. Simulations were undertaken for four types of indoor window covering with different levels of resistance.

Free running and conditioned modes. The setting of different window coverings produced similar patterns in annual thermal performance in both free running and conditioned modes. However, the simulations with four different window coverings showed that there was slightly more sensitivity among the conditioned houses. For instance, all typical houses showed an enhancement in their annual thermal performance when "heavy drape" covering ($R=0.055$) was applied to the windows. The range of enhancement was on average 7.4% in the free running mode and 9% in conditioned mode. It is worth noting that no enhancement in performances could be achieved by adding a pelmet to a heavy drape cover in the Sydney climate, even though this addition increases the effective resistance of window covering ($R = 0.33$) (See Fig. 4.22).

Seasonal performance. The improvement in a house's annual thermal performance from window covering was due to improvement in the house's winter performance, which occurred because of a reduction in heat transmission through the windows. There were only marginal changes in the summer performance of houses. Figures 6.17 and 6.18, Appendix depicts the breakdown of seasonal thermal performances of both house modes in response to the addition of window covering. The addition of drape covering ($R = 0.055$) caused an average of 7.9% enhancement in winter thermal performance in the free running operation mode. This enhancement improved performance by about 4% for houses in the free running mode.

Heavyweight and Lightweight. The effect of window covering on thermal performance was the same whether the house construction was heavyweight or lightweight. However, the addition of drape covers to windows gave 3% greater improvement in the annual thermal performance of HW houses than LW houses.

Double Storey and Single Storey. If the windows are evenly distributed on all sides of a building the effect of window covering on a house's thermal performance

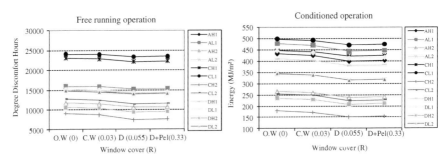

Fig. 4.22 Projected effect of window covering on the annual thermal performances of typical houses in conditioned and free running operation modes

depends on the total window area. The improvement will be greater for a house with a larger window area. Since the window area in typical DS houses is greater than that in typical SS houses, as represented in the study sample, the average improvement in annual thermal performance of the DS houses was found to be 5% greater in both house operation modes.

4.2.2.10 Openable Windows

A building's performance is affected by many features related to windows, such as window size, window frame, distribution of windows in all orientations, window to wall ratios and percentage of openable windows. This last factor seems to be an important aspect of the thermal behavior of a free running house, largely by improving summer behavior. To compare the sensitivity of a house's thermal behavior both in free running and conditioned mode, the application of three different levels of openable windows was simulated.

Free running and conditioned. A house's annual thermal performance could be improved by increasing the percentage of openable windows. An increase from 25 to 75% of the openable window area resulted in an average reduction of 8.7% in annual energy requirement in conditioned mode and 6.7% in annual degree discomfort hours in free running mode (Fig. 4.23). These reductions were due to improvements in their summer performance, which is described in the following.

Seasonal performance. An increase in the percentage of a building's openable windows considerably altered the building's summer performance owing to an increase in natural ventilation. With an increase in the percentage of openable window areas in typical houses from 25 to 75%, without a change the glazed area, their "cooling" degree discomfort hours decreased by an average of 28.5%. The same change decreased their cooling energy requirement by an average of 24.5%. This situation is shown in Figs. 6.19 and 6.20, Appendix.

Winter performances were insensitive to variations in openable windows, unless such variation was connected with changing the window area.

It is clear that the annual improvement was due to the great improvement in the summer performance of houses. However, the summer benefits are almost masked

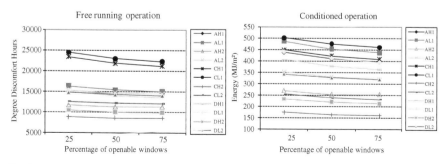

Fig. 4.23 Projected effect of openable window area on the annual thermal performances of typical houses in conditioned and free running operation modes

in annual performances by the dominance of the absolute numbers of the heating energy requirement in conditioned operation mode, and the heating degree hours in free running operation mode.

Contrary to our assumption, there was no significant difference in improvement between the two different house modes. This was because of the nature of the AccuRate software, which automatically counts the beneficial use of natural ventilation to compute the cooling energy requirement of a building in conditioned mode, as described in Sect. 4.1.2.2. However in reality the effect of openable windows on improving the summer performance of a conditioned house depends on the operation of its windows, which in turn depends on the house's occupancy. Depending on how occupants are adapted to indoor temperature, therefore, the role of openable windows in reducing cooling energy requirements might be less than was observed in this study.

Lightweight and Heavyweight. The summer performance of LW houses in both house modes was slightly more sensitive to an increase in the percentage of openable windows than that of HW houses. The potential of a LW construction to change the indoor temperature quickly is accelerated by taking advantage of natural ventilation. By increasing the percentage of openable window area from 25 to 50%, the percentage of improvement in summer performance of the typical houses with LW construction was about 2% more than with HW construction in both house modes.

Single Storey and Double Storey. As noted above, increasing the percentage of openable window area improved the summer performance of SS and DS houses. This improvement was greater in SS than DS houses. It was also greater in their free running mode. By increasing the percentage of openable windows to 50% the summer performance of typical SS houses was improved by 8% more than that of DS houses in free running mode.

The greater improvement in SS houses is related to the situation of the living zone in these houses. With an increase in the percentage of openable windows, the degree of natural ventilation in SS living zones increased more than in the DS living zone. This is because they are also affected by an increase in the percentage of openable windows in the bed zone, while in DS houses living zones and bed zones are separated on two levels. Thus the rate of increase in natural ventilation computed by software, and probably in reality, is greater in SS houses than in DS ones when the percentage of openable windows is increased similarly for both house types.

4.2.2.11 Window to Wall Ratio

The effect of window to wall ratio was simulated in two stages. First the window area in the North (N) and South (S) facades was increased by 15% then 25%, without changing the area of West (W) and East (E) windows. In the second stage a similar increase was applied for W/E windows without changing the area of N/S windows.

Free running and conditioned. In both house modes, increasing the percentage of the window area by 25% made marginal changes in the annual thermal performance. Increasing the window area, whether for N/S or E/W windows, resulted in a slight

4.2 Parametric Sensitivity Analysis of Thermal Performances of Buildings 95

deterioration in the house's performance in conditioned mode. However, the thermal performances improved slightly in free running mode.

Figure 4.24 depicts the situation where an increase in the ratio of N/S windows caused an average 1.1% (0.1–3.3%) enhancement in thermal performance of houses in free running mode, and 1.3% (2.7–0.1%) deterioration in their performance in conditioned mode. The results from changing the percentage of E/W windows are shown in Fig. 4.25. The figures do not make clear the slight changes in the thermal performance because of the large differences in the ranges of absolute numerical value of thermal performance indicators in the 12 samples. For more clarification, the sensitivity of one of the samples (DH2) in response to an increase in the ratio of window to wall area in N/S sides is separately shown in Fig. 4.26.

The reasons for these overall differences between the performances of houses in different modes in response to the same application became apparent when their seasonal performances were examined and this is described in the following.

Seasonal performance. East and west windows influence the summer performance of a building and N & S windows affect the winter performance. The north vertical surfaces receive more irradiation in winter than in summer and east and west vertical surfaces are more influenced by solar radiation in summer than in winter.

The winter performances in conditioned mode were not sensitive to variations in window area, whether the changes were made on the N & S windows or E & W

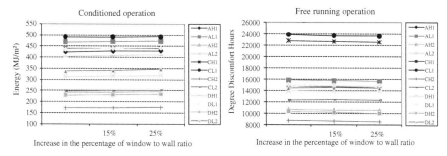

Fig. 4.24 Projected effect of window to wall ratio in north and south orientation on annual thermal performances of typical houses in conditioned and free running operation modes

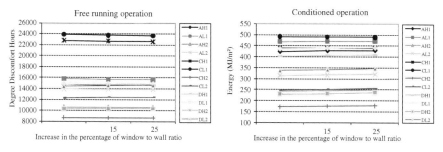

Fig. 4.25 Projected effect of window to wall ratio in east and west orientation on annual thermal performances of typical houses in conditioned and free running operation modes

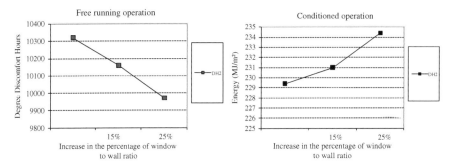

Fig. 4.26 Projected effect of window to wall ratio in north and south orientation on annual thermal performances of a typical house (DH2 sample) in conditioned and free running operation modes

windows. There was only a slight improvement in free running performance of 1.6% on average for an increase in the N & S windows area and 0.9% for the W & E window area.

Houses in both modes benefited from larger window areas in improving their daytime winter performance, since solar radiation enters through the windows and directly heats the building interior. In conditioned mode, however, the amount of artificial heating energy lost through the windows overnight was the same or even greater than the amount of obtained heat. Therefore a smaller window area is appropriate for improving the winter performance of a conditioned house.

As is shown in Figs. 6.21 and 6.24, Appendix, the summer performances of conditioned houses were slightly more sensitive to changes in the window area. When the window areas were changed in both the N & S and E & W orientations, an average increase of 4.2 and 5.1% respectively in the cooling energy requirements of the typical houses was observed.

The effect of increasing the window area on the summer performance of free running houses depended on the house types, and is described in the following.

Single Storey and Double Storey. A comparison between the summer performance of the DS and SS houses in response to an increase in the ratio of windows to walls, demonstrated a considerable difference in the behavior of these two house types. Unlike in DS houses, the summer performance of SS houses, in free running mode, was improved by increasing the window area. For instance a 25% increase in the area of N/S windows resulted in an average of 1.03% improvement in the summer performance of SS houses and of 2.2% degradation in the summer performance of DS houses.

The difference appears to be that in SS houses the improvement in natural ventilation outweighs the increase in overheating hours. Such a pattern was not observed in DS houses.

Larger window areas improve natural ventilation but increase overheating hours during summer time. In this climate zone therefore, the simulation suggests that improvement due to natural ventilation outweighs the penalty of overheating conductive gain and solar loads.

Lightweight and Heavyweight. Heavyweight houses are more affected by changes in the window area. This is due to the effect of thermal mass, which delays heat conduction (transmission), and means that the quality of the indoor environment is maintained for a longer time than with lightweight construction. Increasing the window area in HW houses reduces the effect of the mass by accelerating heat transfer. Since LW houses have this potential characteristic, their sensitivity to changing the window size is less than that in HW houses.

4.2.2.12 Internal Walls

In a typical construction where no additional insulation is considered, the mass of an internal wall is a key issue in improving the thermal behavior of a conditioned zone, particularly when the wall is built between a conditioned and an unconditioned zone. In this situation the role of the internal wall is relatively similar to that of an external wall in improving the thermal conditions of its adjacent spaces. The amount of heat transfer between two rooms depends on the massiveness of the internal wall.

To demonstrate the influence of internal walls on the thermal performance of a house in different modes, typical houses were simulated with four different internal wall constructions and these were compared. Figure 4.27 shows the annual thermal performance of typical houses in free running and conditioned modes, with internal plasterboard as well as with other specified materials (See Sect. 4.1.3 for other specifications)

Free running and conditioned. It was found that the performances in both the conditioned and free running modes were considerably affected by the type of internal wall construction. The effect was greater in free running buildings.

The annual thermal behavior in conditioned mode was enhanced by an average of 8.2% when plasterboard was replaced by brick. This enhancement increased to an average of 19.1% in free running mode. The reasons for this greater improvement become apparent when comparing the houses' seasonal performance as follows.

Seasonal performance. A breakdown of seasonal performance (Figs. 6.25 and 6.26, Appendix) demonstrates that both the summer and winter performances of free

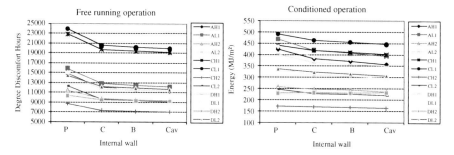

Fig. 4.27 Projected effect of internal wall on the annual thermal performances of typical houses in conditioned and free running operation modes

running houses were more affected by the internal wall mass than were those in the conditioned mode. In the latter an average of 6.4 and 11.3% reduction in the annual heating and cooling energy requirement respectively was observed when plasterboard internal walls were replaced by brick. In free running houses the reduction on average was 20.7 and 15.7%. The greater overall improvement in this case therefore would appear to be the result of a greater improvement in winter performance.

Lightweight and Heavyweight. The thermal performance of a house with LW construction was considerably more affected by an increase in the internal wall mass than when the construction was HW. Indeed this alteration changed the characteristic of a LW house to HW. In some cases, a change in the internal walls of LW houses from plasterboard to brick resulted in quite similar and even better thermal performance than was observed in the HW base models. This situation was particularly observed among typical single storey houses in free running mode. Of course this observation accords with conventional wisdom regarding the relative effectiveness of internal and external thermal mass in a temperate climate.

Single Storey and Double Storey. Brick walls produced a greater improvement in the thermal performance of SS houses than in DS houses in both house modes. This was particularly noticeable for those with LW construction, which achieved indoor thermal quality close to or even better than the thermal quality of HW SS houses with plasterboard internal walls. Figure 4.28 illustrates this situation for a typical house (A1).

4.2.2.13 Infiltration

An increase in the infiltration rate has been found to deteriorate the annual thermal performance of conditioned houses in other studies and in other climates (Willrath, 1997). This study tested the observation for free running houses as well as those in conditioned mode.

For this purpose, three different infiltration rates were simulated. The infiltration rate was increased from 1 to 5 air changes per hour (AC/hr). Indeed 5 AC/hr

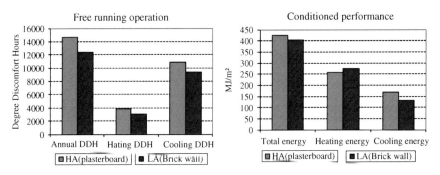

Fig. 4.28 Comparison between the thermal performances of a single storey house (A1) in the Sydney climate when its construction is HW with plasterboard indoor walls, and when its construction is LW with brick internal walls

4.2 Parametric Sensitivity Analysis of Thermal Performances of Buildings

already starts to be ventilation rather than just infiltration; however, the purpose of choosing this parameter is to highlight the significant differences between its effects on the thermal performance of houses when considering its interaction with house operation mode.

Free running and conditioned. Increased infiltration rates degraded the thermal performance of a house in both free running and conditioned modes. This situation is depicted in Fig. 4.29 for all the houses when the infiltration rate increased to 5 air changes per hour.

With this increase, the annual thermal performance of the houses deteriorated by an average of 17.8% in conditioned mode and 14.7% in free running mode. The slightly greater deterioration in the performance of conditioned houses became clear in separating their annual performance into seasonal performances as is shown in the following.

Seasonal performance. The effect of infiltration on the seasonal performance of houses is illustrated in Figs. 6.27 and 6.28, Appendix. Higher infiltration rates significantly degraded the winter performance of the houses in both modes. Changes in the infiltration rate from 0 to 5 air changes per hour caused an average deterioration of 27.2% in their winter performance in conditioned mode, while it was 20.8% for those in free running operation.

The response to increased infiltration rate changes in the summer performance of a house depends on the house mode. By increasing the rate of infiltration (5 air changes per hour), the summer performance for free running houses improved slightly (5.4%), whereas there were no noticeable changes in summer performance in the conditioned mode. Moreover, double storey houses recorded a slight deterioration in their conditioned mode summer performance.

Infiltration causes a latent cooling load in this climate for a conditioned house. By increasing the infiltration rate (5 air changes per hour) in typical houses, the latent cooling increased on average by 32.3%, while "sensible" cooling was reduced by 5%. Depending on the amount of increase in latent cooling energy (MJ/m^2), the total cooling energy requirement of a house can increase by increasing the infiltration rate. The infiltration rate, therefore, is an important parameter in architectural design in a humid summer climate.

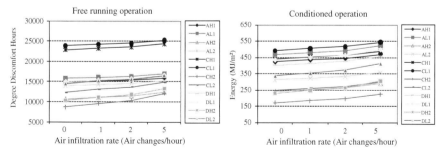

Fig. 4.29 Projected effect of infiltration on the annual thermal performances of typical houses in conditioned and free running operation modes

Heavyweight and Lightweight. The effect of increasing infiltration rates on the thermal performance of the houses with HW construction was greater than on those with LW construction. By increasing the infiltration rate to 5 air changes per hour, deterioration in the thermal performance of HW houses was on average 7% greater than in LW houses in free running mode, and 4% greater than in LW houses in conditioned mode.

4.2.3 Summary of Thermal Performance Analysis

The following conclusions can be drawn from an analysis of the results of the simulations of conditioned and free running houses in the Sydney climate.

- Comparisons between the thermal performances of the typical houses in free running and conditioned mode, demonstrated that the variations in the thermal performance of the houses in response to variations in design features were often markedly dependent on the house operation mode. The free running performance of the houses was clearly different from their performance in conditioned mode. Thus any effort to improve the thermal performance of a house in conditioned mode does not necessarily improve its thermal performance in the free running operation mode. In fact, an effective technique for enhancing the thermal performance of a conditioned house could actually diminish its free running performance. Even if the thermal performance in both free running and conditioned modes of a house is improved, the extent of the improvement is not necessarily the same in both modes. These significant differences between the performances of houses in different operation modes are evidence that the rating of an efficient design for a free running house should be different from that for a conditioned house.
- The effect of changes in a design parameter on the seasonal performance of the houses depended on the house operation mode. Sometimes the seasonal trade-offs were different for each house operation mode. Even if a particular modification in a design parameter produced a similar effect in different operation modes, the relative changes were not the same for both modes.
- A comparison between the thermal performance of houses with LW and HW construction illustrates that the LW houses are sometimes able to achieve a comparable performance to HW houses, particularly when they are in free running mode. LW houses, therefore, could under certain circumstances achieve a more favourable result in a free running rating scheme than in an energy-based rating scheme.
- From all comparisons between the thermal performances of DS and SS houses it was found that the numerical range of annual thermal performances for DS houses was less than that for SS houses, whether in free running or conditioned mode. Figure 4.30 shows this situation in the Sydney climate.

DS houses compared to SS houses, therefore, are most likely to be given better values in a rating system. This is probably due to there being a higher proportion of

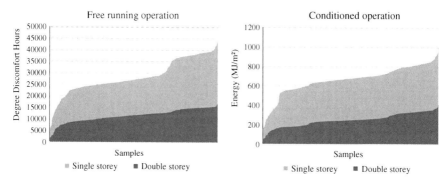

Fig. 4.30 Comparison between the range of numerical values of the thermal performance of DS and SS houses in conditioned and free running modes

envelope in the SS houses for a given volume. In conditioned mode, therefore, the fabric heat flux per unit of floor area is greater in the SS houses.

In free running mode, although the indicator is free of area weighting, DS houses still achieve better grades in an assessment system. This phenomenon in free running mode is no doubt the result of the type of design of DS houses. Placing the bed zone above the living zone considerably reduces the external fabric area for the living zone. Since the thermal performance of a building depends heavily on the performance of its living zone, reducing the external surface of the living zone results in a considerable improvement in the performance of the DS house. However, in a DS design in which the living zone is constructed on the upper floor, this result may be different.

For an accurate evaluation system of the thermal performance of a building, it may be better to separate the score bands for SS and DS houses.

- From the results obtained, the importance of design features can be evaluated according to their potential for improvements in efficiency. The relative strength of each variable affecting the thermal performance of the typical base houses is given in Table 4.6. The percentages for conditioned houses indicate the average annual energy variations, and for free running houses indicate annual degree discomfort hour variations, corresponding to the changes in the parameters from their base case values.

When the building parameters are ranked in order of the percentage of their effect on the thermal performance of the houses in free running mode, this ranking does not correspond with the rank order of the conditioned houses. This situation is shown in Fig. 4.31. The finding demonstrates that the choice of application of any measure to improve a house thermal performance depends on the house operation mode. This will be discussed in further detail in the next section.

It is important to note that while the results show the potential of various design features for improving design efficiency of typical houses, their contribution to

Table 4.6 Percentage variations in annual thermal performance of the typical houses in free running and conditioned modes

Parameters	Percentage thermal performance variations	
	Free running (%)	Conditioned (%)
Ceiling insulation (non to R4.0)	36.4	46.1
Wall insulation (none to R4.0)	7.8	9.7
Floor insulation (none to R2.0)	4.7	1.6
Wall colour (0.30–0.85 absorbance)	3.5	3.6
Roof colour (0.30–0.85 absorbance)	3.16	22.6
Orientation	6	5
Window overhang (non to 1 m)	4.8	3.7
Glazing type (SG Clr to SG Refl.)	4.6	2.1
Window covering (open W. to Drape)	7.4	9
Openable window (0.25–0.75)	6.7	8.7
Window to wall ratio (N & S)	1.1	1.3
Window to wall ratio (E & W)	0.24	1.7
Internal wall (Plasterboard to Brick)	19.1	8.2
Infiltration (0–5 air change per hour)	14.7	17.8

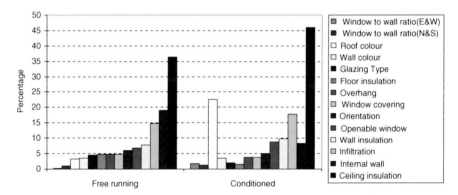

Fig. 4.31 A comparison between the effectiveness of design features in improving thermal performance of the typical houses

actual thermal performance and the relationship among them is not known. A regression analysis is described in the next section to illustrate their actual contribution to changing a house's thermal performance.

4.3 Relationship Between Thermal Performance of Buildings on the Basis of Energy and Thermal Comfort

Significant differences were observed in the parametric comparison between the performances of houses in different operation modes. It is important to identify how the contribution of the various design features differs in improving the thermal

4.3 Relationship Between Thermal Performance of Buildings   103

performance of a house in different operation modes; and how far the different performances of a house are correlated. For this purpose statistical analyses have been employed.

First this analysis investigates a probabilistic correlation between the indicators of house performances in free running and conditioned operation modes. This is followed by an investigation of the contribution of design features to the thermal performance of houses in different operation modes, using multi-regression analysis in order to compare their significance in an efficient design for both a free running and conditioned house.

4.3.1 Correlation Coefficient

The Pearson correlation coefficient was applied to measure the extent to which the thermal performances of the simulated samples in different operation modes are linearly related. It was also used to estimate the strength of the relationship between the thermal performance of houses in conditioned mode and in free running mode in the data set of simulations. Before that, the normality of the variables was checked to ensure the applicability of this coefficient.

The simulated samples that were used for the purpose of parametric sensitivity analysis in this chapter were used again to identify the correlation. The correlation between the thermal performance of conditioned houses on the basis of energy, and that of free running houses on the basis of DDHs (DDHs with area weighting and DDHs without area weighting as described in Sect. 4.1.2.6) is shown here to illustrate how far the addition of area weighting affects the strength of the correlation between the two indicators of energy performance and comfort performance.

Correlation between thermal performances of the typical houses in different modes. The scatter plots of the indicators of the thermal performance of the typical houses (with two different constructions: HW and LW) in conditioned and free running mode for six defined occupancy scenarios are given in Fig. 4.32a–f. The figures show a linear relationship with a relatively strong correlation ($r^2 > 0.5$) between these two indicators for all six occupancy scenarios. However the correlation is not the same for all scenarios; this means that the occupancy scenario is an important parameter in analyzing the thermal performances of a house, and consequently in ranking houses in any related rating system.

It is notable that the correlation is stronger when their free running performance is indicated by degree discomfort hours *without* area weighting.

As discussed before, the evaluation of a house thermal performance in terms of thermal comfort should logically be free of area weighting. This issue was discussed in Sect. 2.4.3 as a shortcoming in the current rating systems and was investigated by the author in Kordjamshidi et al. (2005a). The addition of area weighting to the degree discomfort hours was originally assumed to be necessary to make a better match between the two indicators of a house thermal performance in different modes. However, the finding of this section shows that area weighting is not

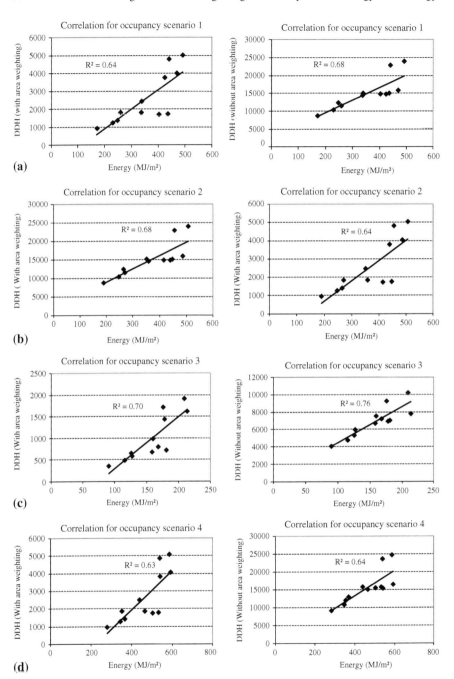

Fig. 4.32 (continued)

4.3 Relationship Between Thermal Performance of Buildings 105

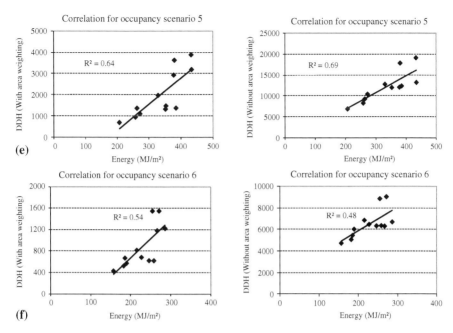

Fig. 4.32 Correlation between thermal performances of the typical houses in different operation modes. (**a**) Correlation between indicators for 1st occupancy scenario. (**b**) Correlation between indicators for 2nd occupancy scenario. (**c**) Correlation between indicators for 3rd occupancy scenario. (**d**) Correlation between indicators for 4th occupancy scenario. (**e**) Correlation between indicators for 5th occupancy scenario. (**f**) Correlation between indicators for 6th occupancy scenario

relevant for this purpose. Therefore the thermal performance of free running houses is reported in degree discomfort hours *without* area weighting.

Correlations between the thermal performances of all the simulated houses in different operation modes. A total number of 620 samples, simulated separately in free running and conditioned mode (1,240 simulations), for the Sydney climate was applied to the data set.

Figure 4.33 demonstrates that the correlation between these indicators is positive and strong ($r^2 = 0.69$). On a bivariate basis, it indicates that 69% of the variation in predicted energy (MJ/m^2) in the Sydney climates can be explained statistically by its relation to DDH. The scatter diagrams in Fig. 4.33 demonstrate the strength of that relationship. Nevertheless, a close observation of the points in the figure suggests that there appear to be at least two or three separate linear clusters of points. Because of this observation, the simulated models were separated into the specific pairs, namely SS/DS and HW/LW.

Earlier a significant difference was observed between the thermal performances of the SS and DS houses in the parametric sensitivity analysis. To clarify the relationships in Fig. 4.33 further, parallel correlation analyses were then conducted for double-storey and single-storey houses (Fig. 4.34).

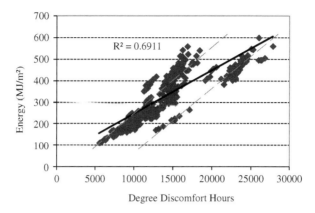

Fig. 4.33 Correlation between the indicators of house thermal performance

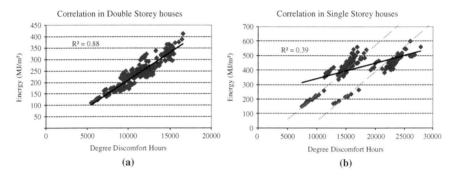

Fig. 4.34 Correlation between the indicators of house thermal performance among DS and SS houses. (**a**) Double storey. (**b**) Single storey

The data points in Fig. 4.34a, which are limited to the double storey cases, describe a much clearer linear relationship between the variables, with $r^2 = 0.88$. The results for single storey cases are equally clear – there are two separate linear clusters of data points. Given the evident spread between those two clusters, it is not surprising that for the single storey cases as a whole the correlation, though still strong, is now ($r^2 = 0.39$).

The strong correlation in DS houses is related to the architectural design of these houses. As noted before, the thermal performance of a house strongly depends on the thermal performance of its living zone, because this zone is occupied for the majority of time in all defined occupancy scenarios. Because the bed zone is generally disposed above the living zone in the DS houses, the external surface area of the living zone in this house type is typically less than that in single storey houses. Therefore the free-running performance of a single storey house is more affected by outdoor climate than that of a DS house. The difference between the free-running and conditioned performances of a single storey house is then greater than that in a double storey house.

4.3 Relationship Between Thermal Performance of Buildings

This observation points to a key difference between the characteristic thermal performance of two storey and single storey houses, and reflects immediately on the reliability of any system which assesses those house types together under a single rating framework.

Figure 4.34b, shows that there appear to be two or three separate linear clusters of points in single storey houses. This observation led to the decision to separate the samples into the two generic house construction forms: "heavyweight" and "lightweight", which are a key variable in the simulation data set. Figures 4.35a, b focus on the single storey cases and describe the impact of the LW versus HW variable on the relationship indicated in the previous scatter plot (in Fig. 4.34a). As it happens, the introduction of the HW/LW variable did nothing to clarify the meaning of the two clusters of linear points that appeared in Fig. 4.35a.

Parallel correlation analyses were then conducted for three groups of single storey houses, which were modelled on the basis of the typical SS models, A1, C1, and D1, as specified in Sect. 4.4. The correlation for all three groups was strong ($r^2 = 0.8$); however, the distribution of points on the scatter plots in Fig. 4.36 shows more than one linear cluster of points in each group. These observations suggest that the effects of the building envelope on the quality of the thermal performance of the building depend on its operation mode. This interpretation, which already was a conclusion in the parametric sensitivity analysis, is taken further in the

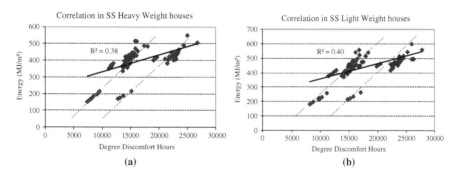

Fig. 4.35 Correlation between the indicators of house thermal performance among SS houses with different construction. (**a**) Heavy weight. (**b**) Light weight

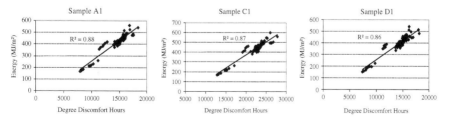

Fig. 4.36 Correlation between the indicators of house thermal performance among SS houses

following section to find the contribution of each parameter to improving the thermal performance of a house in different modes.

4.3.2 Multivariate Regression Analysis

Multivariate regression analysis is one of the most widely used statistical techniques for investigating and modelling the relationship between one variable referred to as a response or dependent variable, and one or more other variables, called predictor or independent variables. It is typically used to identify those variables among a series of predictors that best predict the variation in a dependent variable, and to provide an estimate of how much variation in the dependent variable can be explained by variation in the predictor variables. Applications of regression are numerous in every field, and occur in building performance research whether it is based on experimental or simulated data (Ben-Nakhi and Mahmoud, 2004; Kordjamshidi and King, 2009; Thornton et al., 1997). Also of interest is that regression analysis applied exclusively to simulated data underpins the development of some current rating tools (for example FirstRate, the mandated house energy rating tool in the state of Victoria, Australia, and the regulatory impact studies that support them (Energy Efficient Strategies, 2002)).

A multivariate regression analysis can therefore be applied to a similar study such as this, to determine which parameters of building design features contribute most to overall thermal performance improvement. The main objective of this task is to compare the contribution of the parameters to improving the thermal performance of houses in different operation modes or, in other words, to reducing the annual energy requirement and the degree discomfort hours of houses. For this purpose, even a limited contribution of any parameter is important.

A total number of 1,240 simulations used for parametric sensitivity analysis was undertaken for the following multivariate regression analysis.

One of the most important tasks in regression modelling is the selection of appropriate predictor variables (Beirlant et al., 2005; Gunst and Mason, 1980; Montgomery and Runger, 2002) to be used in subsequently defining the response variable. In most instances it is desirable that the selected variables make physical sense, as well as being useful predictors. Sometimes previous experience or underlying theoretical considerations can help to establish the predictor variables (Montgomery and Runger, 2002).

Fourteen parameters, which are specified in Sect. 4.1.3, Table 4.5, were selected as the main predictors of thermal performance of a house. The likely effect of these parameters was observed from the sensitivity analysis. These parameters are also those identified in other studies as the main fabric building variables which affect the thermal performance of buildings (Willrath, 1997).

It was observed that the main differences between the annual thermal performances of simulated houses were related to three main characteristics of the base models, namely house construction (heavyweight and lightweight), house type (single storey and double storey) and house plan (6 typical house designs). Therefore

4.3 Relationship Between Thermal Performance of Buildings

these three main "design" characteristics of base models were selected as the three parameters to be added to the previous fourteen "fabric" predictors.

To determine the contribution of each of the specified 17 parameters in predicting the thermal performance of a house, the multivariate regression analysis required that all variables be entered into the analysis in a single step. A method which is known as "Enter"[11] procedure was adopted, using SPSS software.

The standardized coefficients correspond to beta weight and a pseudo r^2 statistic is available to summarize the strength of the relationship between the parameters of the design features and indicators of house thermal performance. A standardized coefficient was used in the interpretation, as each parameter was measured in different units.

Using the multiple regression analyses, Table 4.7 indicates how important the 17 variables are as predictors in two contexts: predicting energy (MJ/m^2) for conditioned houses, and predicting DDH for free running houses. This table indicates that the 17 variables (or parameters) do very well in explaining any variation in energy as the dependent variable, where $r^2 = 0.84$ in the Sydney climate.

In contrast, these predictors explain only 53% of the variation in the DDH for free running houses in the Sydney climate. In other words, the same 17 variables do not explain nearly half the variation in DDH for free running houses.

This result indicates a significant difference between a desirable design for conditioned houses and that for free running houses. The amount of unexplained variance for free running houses might be explained by the way occupants operate their homes, in terms of being at home, opening windows and using curtains. However, it certainly needs further research.

In order to investigate how climate is important in designing both a conditioned and a free run house, a similar simulation and regression test was done for Canberra (Australia), which has a moderate climate but with severe winters. When the simulated climate was changed from Sydney to Canberra, the effect of the 17 parameters on predicting the annual energy requirements of conditioned houses changed only by 3% (84–81% = 3%), whereas the effect of these parameters on predicting DDHs for free running houses changed by 13% (53–40% = 13%). This implies that climatic parameters are more important in designing an efficient free running house than they are for an efficient conditioned house. It also can be

Table 4.7 Estimated r^2 from multivariate regression analysis between the annual thermal performances and design features

Climate	Predicted variable	Predictor variables	r^2
Sydney	Conditioned performance (MJ/m^2)	17 design features	0.84
	Free running performance (DDH)	17 design features	0.53

---

[11] There are several selection methods for specifying how independent variables are entered into an analysis. These methods are: enter, stepwise, remove, backward and forward. Enter is a procedure for variable selection in which all variables in a block are entered in a single step.

inferred that free running houses are more sensitive to different climates than are conditioned houses.

This significant effect of climate on the thermal performance of a free running house compared to that on conditioned houses means that if a designer wishes to design a naturally ventilated house rather than an air-conditioned one, a focus on the immediate climate impact is likely to give significantly greater dividends than attention to the fabric of the building. That, however, is not to say that the quality of the fabric cannot help in improving the thermal performance of a free running house. It does mean that different design types require different approaches for the provision of an efficient design.

The relative importance of each of the seventeen building fabric variables (design features) explaining the variations of thermal performance of the simulated houses, is the focus of Table 4.8. Although the values of some of these are not statistically significant, all of the variables have been retained for further analysis as effective parameters for improving the thermal performance of buildings.

Generally a statistically insignificant value for some of the parameters is explained by the range considered for changing those parameters in the typical base cases. For instance, "window to wall ratio" with a significance greater than 0.05 is not statistically an important variable for predicting annual thermal performance of a house. Parametric sensitivity analysis also demonstrated that increasing the proportion of windows in the typical houses by 15 and 25% could only produce about 1% change in the annual thermal performance of those houses. However, a greater increase in the size of windows will result in greater changes in the annual thermal performance of the typical houses, and therefore this parameter could then have significant value in the statistical analysis. Moreover, the effect of this parameter

Table 4.8 Ranking of design features in relation to their importance for the houses' thermal performance (based on standardised regression coefficient)

Free running (degree discomfort hours without area weighting as indicator)				Conditioned mode (energy MJ/m^2 as indicator)			
Rank	Variable	β	Sig.	Rank	Variable	β	Sig.
1	X15 (house type)	0.6	0	1	X15 (House type)	0.749	0
2	X1 (ceiling insulation)	0.265	0	2	X1	0.36	0
3	X16 (house construction)	0.242	0	3	X16	0.274	0
4	X5 (infiltration)	0.084	0.005	4	X5	0.099	0
5	X3 (floor insulation)	0.068	0.025	5	X11	0.091	0
6	X9 (orientation)	0.059	0.056	6	X2	0.079	0
7	X4 (internal wall)	0.049	0.089	7	X17	0.059	0.001
8	X10 (glazing type)	0.046	0.13	8	X9	0.057	0.002
9	X8 (shading device)	0.042	0.165	9	X12	0.039	0.025
10	X17 (house plan)	0.038	0.202	10	X4	0.024	0.16
11	X11 (roof colour)	0.027	0.352	11	X13	0.023	0.194
12	X2 (wall insulation)	0.023	0.457	12	X14	0.023	0.179
13	X6 (windows covering)	0.021	0.483	13	X3	0.018	0.311
14	X14 (windows to wall ratio, E/W)	0.015	0.619	14	X6	0.016	0.343
15	X13 (window to wall ratio, N/S)	0.011	0.711	15	X7	0.01	0.563
16	X12 (wall colour)	0.01	0.734	16	X10	0.01	0.573
17	X7 (openable window)	0.002	0.935	17	X8	0.006	0.722

changes under the interaction effect of other parameters. The absolute size of windows is known to be relevant for the thermal performance of buildings in relation to cooling needs in summer (Persson et al., 2006). It has been identified as an important parameter by experts, home-buyers and stakeholders in a general sense in the development of house ratings, and specifically as a parameter which plays an important role in improving the energy efficiency of buildings. Therefore even design features with low (statistical) significance have not been removed in the following analysis.

A comparison has been made between the importance of the 17 design features in different house modes and on the basis of different indicators. By a considerable margin, the most variation in the thermal performance of the typical houses was related to the house type (X15), namely double storey and single storey houses. This situation can be seen in both conditioned and free running mode operations. This provides evidence of the significant effect of house type in evaluating the thermal performance of a house, which should be taken into consideration in an accurate house performance evaluation system.

In parallel analyses shown in Table 4.8, for both conditioned and free running modes, the most important predictors, in order, were found to be house type (X15), ceiling insulation (X3) and house construction (X17). Beyond that point both the sequence and the statistical significance of the variables (according to their beta coefficients) vary considerably. For instance, roof colour (X11) and wall insulation (X2) are clearly significant in the conditioned mode analysis, but are well down the list and far from statistically significant in the free running houses. Only the multivariate analyses have shown other factors to be more important.

These observations once again have the clear implication that it cannot be assumed that a design for good predicted building performance in conditioned mode achieves good thermal performance in free running mode. A design for conditioned buildings is reasonably related to the building envelope characteristics and to the fabric of the building. Ultimately it relates to those attributes that protect or isolate the building interior from environmental loads, in order to maintain indoor thermal comfort conditions with minimum energy consumption used to overcome those loads. The determinants of free-running performance are more complex, as has long been implied by the alternative terminology "climate responsive". Evidence for this argument was seen before in the parametric sensitivity analysis.

4.4 Conclusion

The results from the parametric sensitivity analysis and multivariate regression analysis effectively demonstrate that the contribution of the 17 design features identified, to the improvement of the thermal performance of a house depends on the house operation mode. Therefore, the choice of application of any measure to improve a house thermal performance depends on the house operation mode

Application priorities for changing design features in order to improve the thermal performance of a house in free running mode are illustrated in Fig. 4.37. This

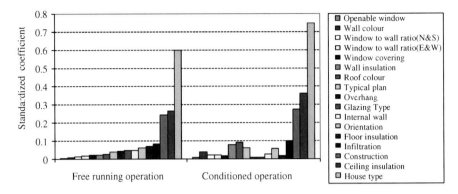

Fig. 4.37 A comparative analyses on the rankings of design feature in different house modes

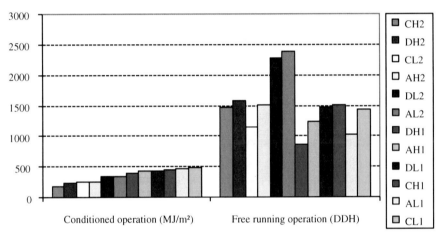

Fig. 4.38 Comparison between ranking typical house performances in conditioned and free running operation modes

figure depicts the ranking of design features on the basis of their strength in changing the thermal performance of the simulated houses in free running operation mode. The ranking is clearly not the same as that for modifying the thermal performance of the same houses in conditioned operation mode.

The result of regression analysis, confirming the results from the parametric sensitivity analysis, demonstrates that design advice inferred from a rating tool to improve the thermal performance of a free running house should differ from that given for a conditioned house. An efficient free running house might not be efficient if it is operated in conditioned mode. This is illustrated in Fig. 4.38. It shows that when the typical houses in conditioned mode were ranked in order, this ranking did not produce the same order as for free running houses.

It is therefore clear that the regulatory framework for rating free-running houses should differ from that for conditioned houses.

References

Akbari, H., Samano, D., Mertol, A., Bauman, F., Kammerud, R.: The effect of variations in convection coefficients on thermal energy storage in buildings part I – interior partition walls. Energy Build. **9**(3), 195–211 (1986)

Al-Homoud, M.S.: Computer-aided building energy analysis techniques. Build. Environ. **36**(4), 421–433 (2000)

ASHRAE: ASNI/ASHRAE Standard 55-2004, Thermal Environmental Conditions for Human Occupancy. American Society of Heating, Refrigeration and Air Conditioning Engineers, Inc, Atlanta, GA (2004)

Auliciems, A., Szokolay, S.V.: Thermal Comfort. PLEA in association with Department of Architecture, University of Queensland, Brisbane (1997)

Australian Bureau of Meteorology: (2006). Accessed 15 Sep 2007, from http://www.bom.gov.au/lam

Ballinger, J.A.: The nationwide house energy rating scheme for Australia (BDP environment design guide no. DES 22). The Royal Australian Institute of Architects, Canberra (1998a)

Ballinger, J.A., Cassell, D.: Solar efficient housing and NatHERS: an important marketing tool. Proceedings of the Annual Conference of the Australian and New Zealand Solar Energy Society, pp. 320–326. Sydney (1994).

Bansal, N.K., Garg, S.N., Kothari, S.: Effect of exterior surface colour on the thermal performance of buildings. Build. Environ. **27**(1), 31–37 (1992)

Beirlant, J., Goegebeur, Y., Teugels, J., Segers, J.: Multivariate extreme value theory. In: Statistics of Extremes: Theory and Applications. Wiley, Chichester (2005)

Ben-Nakhi, A.E., Mahmoud, M.A.: Cooling load prediction for buildings using general regression neural networks. Energy Convers. Manage. **45**(13–14), 2127–2141 (2004)

Bordass, B., Leaman, A.: Occupancy- post-occupancy evaluation. In: Preiser, W.F.E., Vischer, J.C. (eds.) Assessing Building Performance, Elsevier, Sydney (2005)

Cammarata, G., Fichera, A., Marletta, L.: Sensitivity analysis for room thermal response. Int. J. Energy Res. **17**, 709–718 (1993)

Clarke, J.A.: Energy Simulation in Building Design, 2nd ed. Adam Hilger, Oxford and Boston, MA (2001)

Clarke, D.: The Importance of Being Accurate (The Role and Importance of Thermal Modelling in Reducing Energy Consumption in Australian Buildings (No 1)). Association of Building Sustainability Assessors, Surry Hills (2006)

Delsante, A.: Computer User Manual for Program CHEETAH. CSIRO Division of Building Research, Melbourne (1987)

Delsante, A.: A Comparison of CHENATH, the Nationwide House Energy Rating Scheme Simulation Engine, with Measured Test Cell Data, Renewable Energy: The Future in Now, pp. 441–446. Australia and New Zealand Solar Energy Society, Hobart (1995a)

Delsante, A.: Using the Building Energy Simulation Test (Best Test) to Evaluate CHENATH, the Nationwide House Energy Rating Scheme Simulation Engine, Renewable Energy: The Future in Now, pp. 447–453. Australia and New Zealand Solar Energy Society, Hobart (1995b)

Delsante, A.: A Validation of the "AccuRate" Simulation Engine Using BESTEST (no. CMIT(C)-2004-152). CSIRO, Canberra (2004)

Delsante, A.: Is the New Generation of Building Energy Rating Software up to the Task? – A Review of AccuRate, ABCB Conference, Building Australia's Future 2005. CSIRO Manufacturing and Infrastructure Technology, Surfers Paradise (2005)

Drysdale, J.W.: Designing Houses for Australian Climates, 3rd ed. Australian Government Publishing Service, Canberra (1975)

Energy Efficient Strategies: Comparative Cost Benefit Study of Energy Efficiency Measures of Class 1 Buildings and High Rise Apartments in Victoria (Final report for the Sustainable Energy Authority of Victoria). Melbourne (2002)

Fisette, P. (2003). Windows: Understanding Energy Efficient Performance. Accessed 6 June 2008 from http://www.umass.edu/bmatwt/publications/articles/windows_understanding_energy_efficient_performance.html

Foster, R.: Setting Occupancy Factors for Thermal Performance Modelling of Australian Households. Energy Efficient Strategies, Victoria (2006)

Givoni, B.: Man, Climate and Architecture. Applied Science Publishers Ltd, London (1976)

Gunst, R.F., Mason, R.L.: Regression Analysis and Its Application: A Data Oriented Approach. Marcel Dekker, New York, NY (1980)

Hanby, V.I.: Error estimation in bin method energy calculations. Appl. Energy. **52**(1), 35–45 (1995)

Hong, T., Chou, S.K., Bong, T.Y.: Building simulation: an overview of developments and information sources. Build. Environ. **35**(4), 347–361 (2000)

Hyde, R.: Climatic design: a study of housing in the hot humid tropics. The Proceedings of the Australian and New Zealand Energy Society, Darwin (1996)

Hyde, R.: Climate Responsive Design: A Study of Buildings in Moderate and Hot Humid Climates. E & FN Spon, New York, NY (2000)

Isaacs, T.: Revision of the Energy Efficiency Provisions for Housing to Better Allow for the Impact of Ventilation. ABCB, Canberra (2004)

Isaacs, T.: AccuRate: 2nd Generation Nationwide House Energy Rating software. The Royal Australian Institute of Architects, Canberra (2005)

Klainsek, J.C.: Glazing and its influence on building energy behaviour. Renewable Energy. **1**(3–4), 441–448 (1991)

Klein, S.A.: Computer in the design of passive solar systems. Passive Solar J. **2**(1), 57–74 (1983)

Kordjamshidi, M., Khodakarami, J., Nasrollahi, N.: Occupancy scenarios and the evaluation of thermal performances of buildings. Proceeding of ANZSES conference, Townsville, Australia (2009)

Kordjamshidi, M., King, S.: Overcoming problems in house energy ratings in temperate climates: a proposed new rating framework. Energy Build. J. **41**(1), 125–132 (2009)

Kordjamshidi, M., King, S., Prasad, D.: An alternative basis for a home energy rating scheme (HERS). Proceedings of PLEA, Environmental Sustainability: The Challenge of Awareness in Developing Societies, pp. 909–914. Lebanon (2005a)

Kordjamshidi, M., King, S., Prasad, D.: Towards the development of a home rating scheme for free running buildings. Proceedings of ANZSES, Renewable Energy for a Sustainable Future – A Challenge for A Post Carbon World. Dunedin University, New Zealand (2005b)

Lam, J.C., Hui, S.C.M.: Sensitivity analysis of energy performance of office buildings. Build. Environ. **31**(1), 27–39 (1996)

Lee, T., Snow, M.: The Australian climate data bank project. Proceedings of the IBPSA Australia 2006 Conference: Investigating the Roles and Challenges of Building Performance Simulation in Achieving a Sustainable Built Environment, The University of Adelaide, Adelaide (2006)

Littler, J.G.F.: Overview of some available models for passive solar design. Comput. Aided Des. **14**(1), 15–19 (1982)

Lomas, K.J., Eppel, H.: Sensitivity analysis techniques for building thermal simulation programs. Energy Build. **19**(1), 21–44 (1992)

Markus, T.A., Morris, E.N.: Building Climate and Energy. Pitman, London (1980)

Montgomery, D.C., Runger, G.C.: Applied Statistics and Probability for Engineers, 3rd edn. Wiley, New York, NY (2002)

Nielsen, T.R., Duer, K., Svendsen, S.: Energy performance of glazings and windows. Solar Energy. **69**(Supplement 6), 137–143 (2001)

Offiong, A., Ukpoho, A.U.: External window shading treatment effects on internal environmental temperature of buildings. Renewable Energy. **29**(14), 2153–2165 (2004)

Olgyay, V.: Design With Climate: Bioclimatic Approach to Architectural Regionalism. Princeton University Press, Princeton, NJ (1963)

Omar, E.A., Al-Ragom, F.: On the effect of glazing and code compliance. Appl. Energy. **71**(2), 75–86 (2002)

References

Persson, M.-L., Roos, A., Wall, M.: Influence of window size on the energy balance of low energy houses. Energy Build. **38**(3), 181–188 (2006)

Planning: (2006). New 5 Star Requirements: Making Your Home More Energy Efficient. Government of South Australia. Accessed 7 Oct 2006, from www.planning.sa.gov.au

Preiser, W.F.E.: Building performance assessment – from POE to BPE, a personal perspective. Archit. Sci. Rev. **48**(3), 201–204 (2005)

Preiser, W.F.E., Vischer, J.C.: The evolution of building performance evaluation: an introduction. In: Preiser, W.F.E., Visscher, J.C. (eds.) Assessing Building Performance, pp. 3–13. Elsevier, Oxford, UK (2005)

SOLARCH: Project Homes: House Energy Rating, New South Wales Industry Impact Study (A report prepared for the Sustainable Energy Development Authority): University New South Wales (2000)

Shariah, A., Shalabi, B., Rousan, A., Tashtoush, B.: Effects of absorptance of external surfaces on heating and cooling loads of residential buildings in jordan. Energy Convers. Manage. **39**(3–4), 273–284 (1998)

Sowell, E.F., Hittle, D.C.: Evolution of building energy simulation methodology. ASHRAE Trans. **101**(P.1), 850–855 (1995)

Sutherland, J.W.: The solution of psychometric problems using a digital computer. Air. Cond. Heating. **25**(4), 43–49 (1971)

Szokolay, S.V.: Handbook of Architectural Technology. Van Nostrand Reinhold, New York, NY (1991)

Szokolay, S.V.: Introduction to Architectural Science: The Basis of Sustainable Design. Architectural Press, Oxford (2004)

Tarantola, S., Saltelli, A.: SAMO 2001: methodological advances and innovative applications of sensitivity analysis. Reliab. Eng. Syst. Saf. **79**(2), 121–122 (2003)

Tavares, P.F.A.F., Martins, A.M.O.G.: Energy efficient building design using sensitivity analysis – a case study. Energy Build. **39**(1), 23–31 (2007)

Thornton, S.B., Nair, S.S., Mistry, S.I.: Sensitivity analysis for building thermal loads. ASHRE Trans. **103**, 165–175 (1997)

Tuhus- Dubrow, D., Krarti, M.: Genetic-algorithm based approach to optimize building envelope design for residential buildings. Build Environ. **45**(7), 1574–1581 (2010)

Walsh, P.J., Gurr, T.A.: A Comparison of the Thermal Performance of Heavyweight and Lightweight Construction in Australian Dwellings (no. TP44). CSIRO Division of Building Research, Australia (1982)

Williamson, T., Riordan, P.: Thermostat strategies for discretionary heating and cooling of dwellings in temperate climates. Proceeding of 5th IBPSA Building simulation Conference, pp. 1–8. International Building Performance Simulation Association, Prague (1997)

Willrath, H.: Thermal sensitivity of Australian houses to variations in building parameters, 35th Annual Conference of the Australian and New Zealand Solar Energy Society, Canberra (1997)

Zhai, Z.J., Chen, Q.Y.: Sensitivity analysis and application guides for integrated building energy and CFD simulation. Energy Build. **38**(9), 1060–1068 (2006)

Chapter 5
Assembling a House Energy Ratings (HER) and House Free Running Ratings (HFR) Scheme

This chapter describes the proposal for a suitable house rating scheme. A framework for producing an aggregation of free running and conditioned rating schemes is put forward. The reliability of this proposed rating framework is supported by a demonstration of its theoretical sensitivity in improving efficient design quality.

5.1 Rating Building Thermal Performance

A house rating system scores a house by comparing its thermal performance with that of other houses, which are given the same conditions of climate, user behavior patterns and house operation. An accurate house rating system should not discriminate against any type of house design. Previous chapters, however, have reached the conclusion that the current energy-based rating schemes, as exemplified by AccuRate, are likely to discriminate against single storey houses as opposed to double storey houses, and against free running houses in contrast to conditioned ones. The energy efficiency of different house types is in fact not comparable in terms simply of an energy performance index. A large single storey house, for example, is not less efficient than a small double storey house, even though it may have a grade or an energy performance index more than three times greater on average. To avoid such discrimination the following steps should be considered for rating schemes:

- Separating free running houses from conditioned houses
- Separating double storey houses from single storey houses
- Determining the score boundaries for each group separately
- Aggregating the score bands of a house for its free running and conditioned performances to produce one score for the final evaluation of the thermal performance of the house

5.1.1 How Should Building Thermal Performance Bands Be Defined for Rating?

Building regulations usually set the minimum overall requirements for the energy performance index. There are two different approaches for the determination of limits: fixed and customized (Lombard et al., 2009). In the *fixed limit* option, the limit is determined on the basis of regulations, and the threshold value depends on certain parameters such as climate, buildings type and building operation mode, whose impact is to be neutralized. The *customized limit* can be obtained by self-reference, where the threshold value is set by a reference building with at least similar location, climate and building type, but different envelope and system.

There is no single standard method for defining performance band ratings based on five stars, such as the NatHERS, or on ten stars, such as AccuRate (see Chap. 2). The choice of procedure used for determining the boundaries of each star-scale varies from place to place, depending on the particular legislation involved in the promotion of energy efficiency. However, for our purposes the star categories provide sufficient and meaningful differentiation between the relative efficiencies of different houses in the same condition and operation mode.

To establish score bands for the ratings in this study, five separate efficiency categories were first adopted separately for free running (HFR) and conditioned houses (HER), with a range of scores represented by stars. Then, in order to combine the two rating systems, a ten star rating scheme was adopted. The ten categories differentiate sufficiently between the efficiency of the houses' architectural design in both their free running and conditioned performances.

One method used for specifying the star bands is based on the theory of educational measurement and evaluation. The five letter system (A, B, C, D, and F) is commonly used in education and attempts to classify individuals in terms of their performance. There are two types of measurements related to specific standards, namely "criterion- referenced" and "norm- referenced" grading (Ebel and Frisbie, 1991). The second type is used here because of the limited number of data we have from the simulations in Chap. 4.

Norm-referenced measurement is based on relative standards. The purpose of a norm-referenced instrument is to compare the performance of a character with the performance of other characters. The scale is usually anchored in the middle of some average level of performance for a specific set of characters. The units on the scale represent the distribution of performances above and below the average level. In this case, norm-referenced grading was adopted for the range of available data to develop HRS score bands.

"Grading on a curve" is a common technique for grading performance in norm-referenced measurement. It is based on frequency distribution, in which the curve represents a normal distribution. The categories should represent equal intervals on the score scale. Mean and standard deviations of the distribution of performance scores are generally used for determining the range of each category. The range of the normal curve can then be divided into equal segments of standard deviation, according to the number of expected categories.

5.1 Rating Building Thermal Performance

In adopting the *grading on the curve* technique, the 1,240 estimated values of thermal performance of simulated houses in different operation modes, which are representative of house thermal performances in the Sydney climate, have been used. As previously noted, the process has been staged separately for free running houses and then for conditioned houses. The score bands have also been determined separately for single storey and double storey houses in each operation mode.

Figures 5.1a, b and 5.2a, b show the frequency distribution of the annual energy requirements of conditioned houses and of the annual degree discomfort hours of

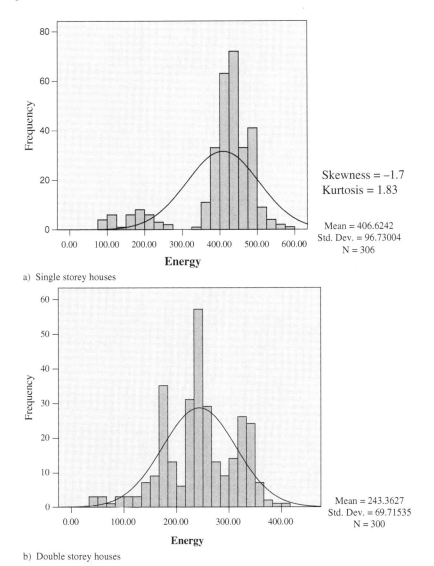

a) Single storey houses

Skewness = –1.7
Kurtosis = 1.83

Mean = 406.6242
Std. Dev. = 96.73004
N = 306

b) Double storey houses

Mean = 243.3627
Std. Dev. = 69.71535
N = 300

Fig. 5.1 Distribution of estimated annual energy requirements in the Sydney climate, **a** Single storey houses, **b** Double storey houses

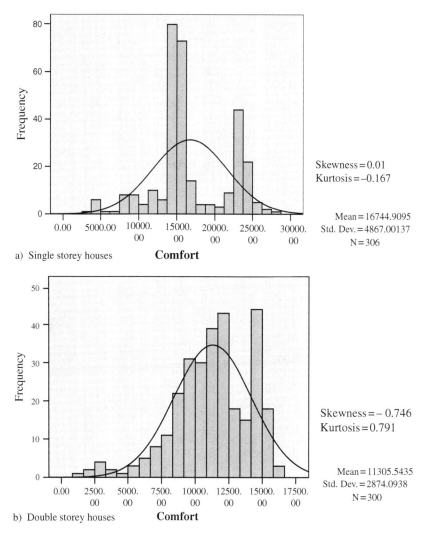

Fig. 5.2 Distribution of annual degree discomfort hours, (**a**) Single story houses, (**b**) Double storey houses

free running houses. The distributions seem to be normal, although in terms of skewness[1] and kurtosis[2] it can be seen that they are not in fact completely normal. However, as these deviations are relatively slight, the assumption of normal distribution is accepted.

---

[1] Skewness is a measure of symmetry.

[2] Kurtosis is a measure of whether the data are peaked or flat relative to a normal distribution.

5.1 Rating Building Thermal Performance

The range of energy requirements and degree discomfort hours is dealt with separately in defining star bands for single storey and double storey houses, with star bands for conditioned houses being defined on the basis of the mean of annual energy requirements and standard deviation for each group, while the star bands for these houses in the free running mode are defined on the basis of the mean of annual degree discomfort hours and standard deviation of the related group. The proposed range of energy requirements and degree discomfort hours for each star rating is presented in Tables 5.1 and 5.2.

The star bands for each group are determined on the basis of the standard deviation and mean values related to that group.

Table 5.1 Proposed range of energy requirements and degree discomfort hours for each star rating for single storey houses in Sydney

Degree discomfort hours (DDH)	Energy requirement (MJ/m^2)	Rating of energy performance	Range on normal curve (from mean)
5,794 or less	188 or less	5 Star	−2.25 Std dev. or lower
5,794<DDH≤ 8,227.5	188<E≤ 237	4.5 Star	−2.25 to −1.75 Std dev.
8,227.5<DDH≤ 10,661	237<E≤ 285	4 Star	−1.75 to −1.25 Std dev.
10,661<DDH≤ 13,094.5	285<E≤ 334	3.5 Star	−1.25 to −0.75 Std dev.
13,094.5<DDH≤ 15,528	334<E≤ 382	3 Star	−0.75 to −0.25 Std dev.
15,528<DDH≤ 17,961.5	382<E≤ 430	2.5 Star	−0.25 to +0.25 Std dev.
17,961.5<DDH≤ 20,395	430<E≤ 479	2 Star	+0.25 to +0.75 Std dev.
20,395<DDH≤ 22,828.5	479<E≤ 527	1.5 Star	+0.75 to +1.25 Std dev.
22,828.5<DDH≤ 25,262	527.5<E≤ 576	1 Star	+1.25 to +1.75 Std dev.
25,262<DDH≤ 27,695.5	576<E≤ 624	0.5 Star	+1.75 to +2.25 Std dev.
More than 27,695.5	More than 624	0 Star	+2.25 Std dev. or higher
Mean = 16,744.9095	Mean = 406.6242		
Std. Dev.= 4,867.00137	Std.Dev.= 96.73004		

Table 5.2 Proposed range of energy requirements and degree discomfort hours for each star rating for double storey houses in Sydney

Degree discomfort hours (DDH)	Energy requirement (MJ/m^2)	Rating of energy performance	Range on normal curve (from mean)
3,402 or less	86.8 or less	5 Star	−2.25 Std dev. or lower
4,839<DDH≤ 6,276	86<E≤ 121	4.5 Star	−2.25 to −1.75 Std dev.
6,276<DDH≤ 7,713	121<E≤ 156	4 Star	−1.75 to −1.25 Std dev.
7,713<DDH≤ 9,150	156<E≤ 191	3.5 Star	−1.25 to −0.75 Std dev.
9,150<DDH≤ 10,587	191<E≤ 226	3 Star	−0.75 to −0.25 Std dev.
10,587<DDH≤ 12,024	226<E≤ 261	2.5 Star	−0.25 to +0.25 Std dev.
12,024<DDH≤ 13,461	2,601<E≤ 296	2.Star	+0.25 to +0.75 Std dev.
13,461<DDH≤ 14,898	296<E≤ 331	1.5 Star	+0.75 to +1.25 Std dev.
14,898<DDH≤ 16,335	331<E≤ 366	1 Star	+1.25 to +1.75 Std dev.
16,335<DDH≤ 17,772	366<E≤ 400	0.5 Star	+1.75 to +2.25 Std dev.
More than 17,772	More than 400	0 Star	+2.25 Std dev. or higher
Mean = 11,305.5435	Mean = 243.3627		
Std. Dev.= 2,874.0938	Std. Dev.= 69.71535		

Ideally, the score of a house performance should be represented by one indicator only, in order to simplify the condition for comparing and rating buildings, and this is found in international rating systems. However, as it was found (see Chap. 4) that performances differed between conditioned and free-running houses, two separate scores have been computed for HFRS and HERS. In order to reduce this to one single score for the architectural design of a house, a combination of the two ratings is desirable. The proposed technique for combining the two rating systems is the aggregation of the scores of free running and conditioned performances of a house as described below.

5.2 The Combination of Two Rating Systems

A new star band can be obtained by simply adding the scores obtained from the free running and conditioned performances of a house. However, such a simple aggregation will not differentiate between the values of efficient design for each house mode. For instance, if a house achieves 5 stars in its free running rating and 2 stars in its conditioned rating, its final score (7 stars) would be similar to that of a building with 5 stars in its conditioned rating and 2 stars in its free running rating. However, these two houses should not achieve a similar score, since their design efficiency and their characteristics are different. If the former was operating as a conditioned house, it would result in more energy consumption for space heating and cooling than the latter.

An algorithm is employed in the following for aggregating the two rating systems to give more value to either an efficient free running house or an efficient conditioned house. Either of these would be rewarded, depending on the policy for reducing energy requirements in the building sector. For instance, where the climate is suitable for taking the most advantage of the outdoor environment, free running houses should have priority for promoting efficient architectural design. This situation would be reversed for promoting efficient conditioned houses if there is a reason for not constructing free running houses.

The method would be flexible enough to make it also applicable for all regions, and it should therefore be applicable and adjustable for national or even international use for HRS.

However, this book is concerned specifically with promoting efficient free running houses. Thus it aims to give more value to such houses in a moderate climate (such as Sydney's), as a matter of policy, in the context of sustainable development. It is intended to encourage the public to adopt free running houses, in order to reduce energy consumption for space heating and cooling.

The proposed algorithm for this purpose is presented in Fig. 5.3. A 10 star rating scheme is proposed for aggregating the scores of a 5 star free running and 5 star conditioned rating. The algorithm is most conveniently executed as a "lookup table",

5.2 The Combination of Two Rating Systems

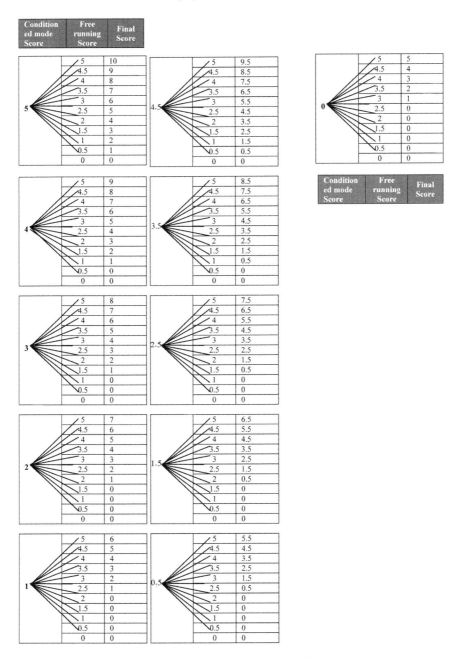

Fig. 5.3 An algorithm for aggregating the scores of HERS and HFRS

as illustrated in Fig. 5.3. The 11 cells of the table correspond to the initial range of possible "half star" ratings for conditioned performance. In each cell of the table:

- the first column allocates the probabilistic score for the conditioned performance of a house in a 5 star rating;
- the second column allocates all probabilistic scores for the free running performance of the house in a 5 star rating;
- the third column allocates the final aggregated score in which:
  - the top score is obtained from adding the conditioned rating score to the 5 star rating of the house in free running performance,
  - the other final scores allocated in the third column are produced by taking 1 score away from the resulting upper final score. This establishes a different value to the different performances of a house by rewarding the free running performance of the house.

In this sort of aggregation the final score is mostly less than the score produced by simply adding the scores of both ratings. For instance in the first group, by adding the first two grades of free running (5) and conditioned (5) ratings, the result will be 10, which is the top star rating in the new rating system. The addition of the second grades in this group, free running (5) and conditioned (4.5), gives a sum of 9.5 but a score of 9 (10−1=9) stars is given in the proposed method of aggregation. Likewise, the third grade in the same group produces 8 stars (9−1=8) for aggregating: (5) stars for conditioned performance and (4) stars for the free running performance of the house.

A regression analysis has been employed in this case to examine the sensitivity of the final score (as a dependent variable) in relation to the free running and conditioned scores (as independent variables).

The regression obtained (5.1) confirms the greater dependency of the final score on free running performance. The standardized coefficient for free running performance is 0.87, where that for the conditioned performance is 0.4.

$$S = 0.40 S_1 + 0.87 S_2 \tag{5.1}$$

In other words, to produce the final score for the thermal performance of the house, the proposed new HRS gives more value to the free running performance, to the extent that the effect of the free running score is more than double the effect of the conditioned score. As a result, a designer would be likely to give priority to free running design for improving the thermal performance of the house in order to achieve an acceptable score in the HRS. This supports the objective of the rating framework – to encourage the public and architects to opt for free running houses.

However, depending on the appropriate policy for reducing energy consumption, the method can also be applied to give more value to conditioned houses, simply by transforming the method of aggregation of the two ratings. In its simplest form, where the conditioned and free running scores are transposed in the lookup table, the higher coefficient (0.87) would apply to the score for the conditioned performance of a house.

5.3 How the New Combined System Evaluates Efficiency

A minimum star level is generally required for designation of efficiency in an architectural design. This varies according to the state or territory, and is proposed by authorities in each jurisdiction. For example, the initial regulatory framework of NatHERS in Australia established a maximum of 5 stars for HERS, with 3.5 stars as the prerequisite for an efficient house design. This is now superseded by the 10 star system of AccuRate, which requires 5 stars as a prerequisite for an efficient design in those states of Australia in which AccuRate is approved as a mandated tool.

Since a 10 star rating is proposed for HRS, with aggregates of 5 stars for HERS and HFRS, the minimum requirement for an efficient design would be 6.5 stars, by which a house would achieve at least 4 stars for its free running performance. As observed above, a house with 4 stars in free running mode would achieve 3.5 stars in conditioned operation mode. The combination of 4 and 3 stars is 6.5, which is thus defined as the minimum requirement for energy efficient design.

Figure 5.4 illustrates the position of acceptable states in the proposed framework in conjunction with stars taken from HFRS and HERS. The final rating score is displayed as the size of the bubble marker. The highest score is 10 and its position is indicated in the Fig. 5.4

The proposed prototype rating framework can be summarized as follows:

- single storey houses are separated from double storey houses;
- the evaluation is based on the house performance in both free running and conditioned mode;
- a house is given two separate scores in a 5 star rating scheme for its free running and conditioned performances;

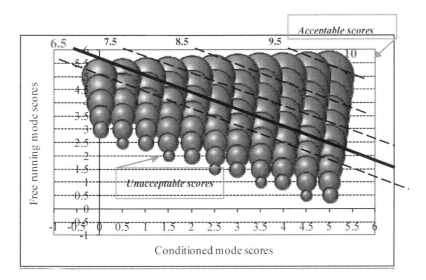

Fig. 5.4 Pictorial expression of the proposed framework

- the scores of free running and conditioned performances are combined under a developed algorithm to produce a final score in a 10 star range;
- the proposed algorithm gives more reward to the free running than the conditioned performance of a house; however it is flexible enough to be inverted, depending on the policy;
- achieving 6.5 stars is the main criterion for acceptable house design, for which a house would get at least 4 stars from HFRS.

5.4 Reliability of the New Rating System

One way to test the reliability of the proposed framework is to test it on real houses by employing it for improving their thermal performance. Another way is testing the framework for its theoretical sensitivity to improvements in efficient design. A simulated improvement in response to the composed rating was in this case employed to determine if the framework delivers expected sensitivity. For this purpose the following steps were taken:

- The design quality of the typical houses was improved by modifying design features in order to enhance first the free running performance, then the conditioned performance of the houses.
- The resulting annual energy requirements and DDHs were input in a regression model to determine the correlation between the indicators of free running and conditioned performances for these "improved" houses.
- All simulated houses were scored on the basis of the proposed rating scheme to check how their scores on the HERS changed in relation to the changes in their score on the free running component.

The linear regression shows a strong correlation between the annual energy requirements and degree discomfort hours of houses when their thermal performances were improved. The scatter plot of this situation is depicted in Fig. 5.5. The scatter plots for "improved" houses are circled in this figure to draw specific attention to them. A strong linear relationship between the thermal performances of "improved" houses is obvious. The correlation is 0.86.

Before the improvement the correlations were 0.69 for the Sydney climate, and 0.56 for the Canberra climate (see Sect. 4.3.1, Fig. 4.33).

This observation indicates that an efficient house design can result in good performances for a house in both operation modes if its design quality is improved greatly for either its free running or conditioned operation modes.

This logical implication needs to be checked against the proposed score bands to see the relationship between the scores of houses in free running (HFRS) and conditioned ratings (HERS). All simulated samples therefore were scored on the basis of the proposed rating. The relationship between the scores is shown in Fig. 5.6. It can be seen that there is a linear relationship between the scores of double storey houses

5.4 Reliability of the New Rating System

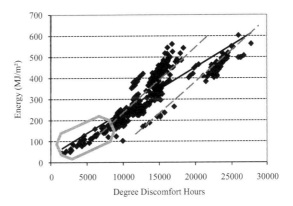

Fig. 5.5 Correlation between energy requirements and degree discomfort hours ($r^2=0.86$ for *area circled*)

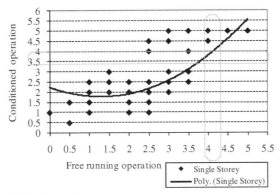

a) Single storey houses

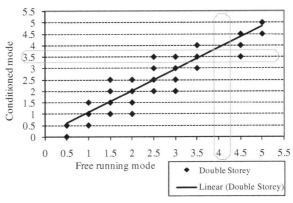

b) Double storey houses

Fig. 5.6 Correlation between the score of houses in free running and conditioned modes

in different modes. However, the relationship is polynomial among the single storey houses. This is evidence again of a significant difference between the characteristics of SS and DS houses, which requires that a rating system separates these two house types for evaluation.

A remarkable point in this approach is that houses with 4 stars in free running mode get at least 3.5 stars in their conditioned performance, but houses with 4 stars in their conditioned performance do not necessarily get a score higher than 3 stars in their free running performance. In other words, an efficient design for a free running house could improve the performance of that house in the conditioned mode if its free running score is not less than 4. This condition has been included in the proposed prototype framework for HRS.

Under these circumstances it appears that the proposed aggregation of HFRS and HERS is an appropriate response to the objective of HRS. It does not compromise the value of the conditioned performance of houses, while highlighting the value of free running houses, in which energy requirements for space heating and cooling will be significantly reduced.

5.5 Conclusion

This book was intended to develop a method for HRS that would give appropriate value to the free running performance of houses, which is missing in the current house rating schemes, so as to encourage the adoption of such houses. In order to achieve this, an aggregation technique was developed to differentiate the value of a free running and a conditioned performance of a house, in which the efficiency of a designed house is evaluated for grading on the basis of its performances in both the modes, and which awards more value to free running performance, but which in application also generally improves conditioned performance. However, the proposed system remains flexible enough to give higher value to either free running or conditioned performances, which therefore makes it adjustable for the promotion of any kind of efficient architectural design, depending on the relevant policy for reducing energy consumption.

The proposed HRS modifies the current rating scheme by introducing the following changes:

- It separates double storey houses from single storey houses to remove discrimination in favour of the value of SS against DS houses.
- It rates the house thermal performance separately for its free running and conditioned operation, then aggregates the star bands to produce a single star rating for final comparative evaluation.

It establishes the criterion (in the number of stars) for assessing an efficient architectural design in order to ascertain the efficiency of house design in *both* operations

References

Ebel, R.L., Frisbie, D.A.: Essentials of Educational Measurement (5th ed.). Prentice-Hall, Inc., Englewood Cliffs, NJ (1991)

Lombard, L., Ortiz, J., Gonzalez, R., Maestre, I.R.: A review of benchmarking, rating and labeling concepts within the framework of building energy certification scheme. Energy Build. **41**(3), 272–278 (2009)

Chapter 6
Appendix

6.1 The Effect of House Envelope Parameters on the Seasonal Performance of Houses in Different Operation Modes

This Appendix reports in full the additional Figures referred in Chap. 4, being the sensitivity of seasonal, winter and summer, performances of simulated houses in response to changing design parameters for two house operation modes in the Sydney climate

- Ceiling insulation

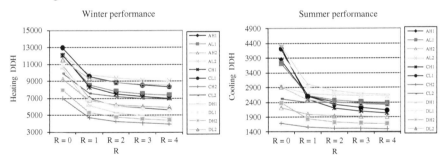

Fig. 6.1 Projected effect of ceiling insulation on the seasonal performance of the typical houses in free running operation mode

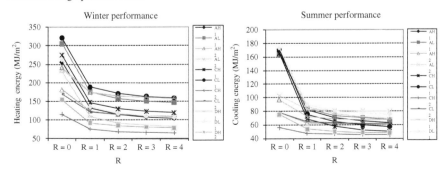

Fig. 6.2 Projected effect of ceiling insulation on the seasonal performance of the typical houses in conditioned mode

- Wall insulation

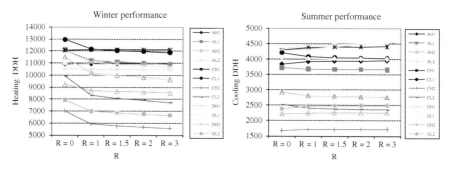

Fig. 6.3 Projected effect of wall insulation on the seasonal performance of the typical houses in free running operation mode

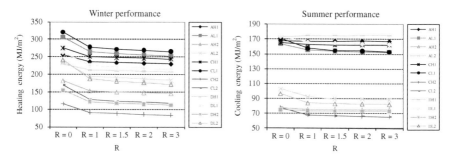

Fig. 6.4 Projected effect of wall insulation on the seasonal performance of the typical houses in conditioned mode

- Floor insulation

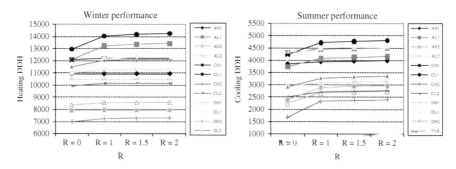

Fig. 6.5 Projected effect of floor insulation on the seasonal performance of the typical houses in free running operation mode

6.1 The Effect of House Envelope Parameters on the Seasonal Performance   133

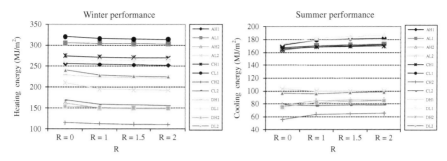

Fig. 6.6 Projected effect of floor insulation on the seasonal performance of the typical houses in conditioned mode

- Wall colour

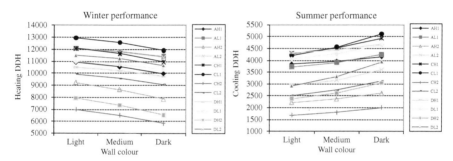

Fig. 6.7 Projected effect of external wall colour on the seasonal performance of the typical houses in free running operation mode

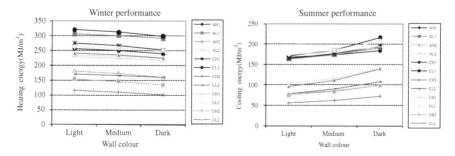

Fig. 6.8 Projected effect of external wall colour on the seasonal performance of the typical houses in conditioned mode

- Roof colour

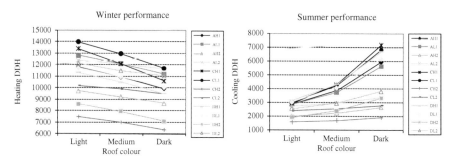

Fig. 6.9 Projected effect of roof colour on the seasonal performance of the typical houses in free running operation mode

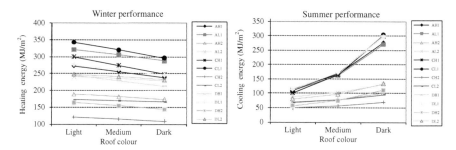

Fig. 6.10 Projected effect of roof colour on the seasonal performance of the typical houses in conditioned mode

- Orientation

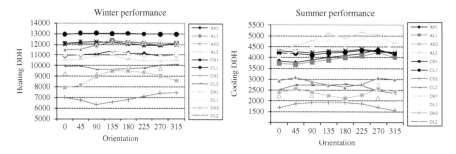

Fig. 6.11 Projected effect of orientation on the seasonal performance of the typical houses in free running operation mode

6.1 The Effect of House Envelope Parameters on the Seasonal Performance 135

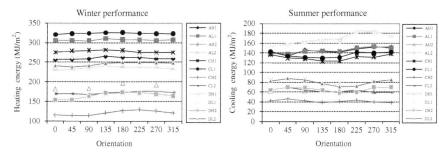

Fig. 6.12 Projected effect of orientation on the seasonal performance of the typical houses in conditioned mode

- Overhang

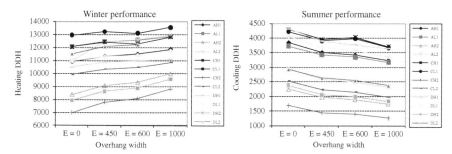

Fig. 6.13 Projected effect of overhang width on the seasonal performance of the typical houses in free running operation mode

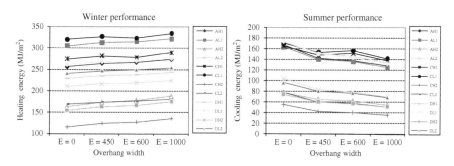

Fig. 6.14 Projected effect of overhang width on the seasonal performance of the typical houses in conditioned mode

- Glazing type

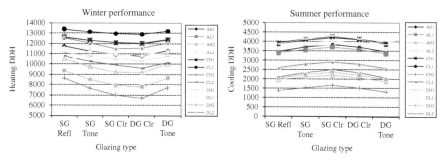

Fig. 6.15 Projected effect of glazing type on the seasonal performance of the typical houses in free running operation mode

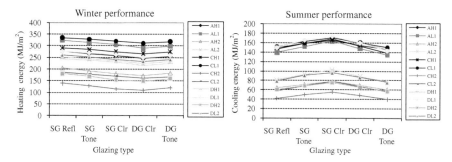

Fig. 6.16 Projected effect of glazing type on the seasonal performance of the typical houses in conditioned mode

- Window covering

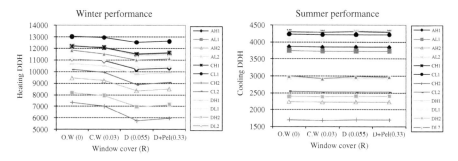

Fig. 6.17 Projected effect of window covering on the seasonal performance of the typical houses in free running operation mode

6.1 The Effect of House Envelope Parameters on the Seasonal Performance   137

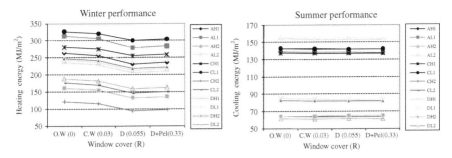

Fig. 6.18 Projected effect of window covering on the seasonal performance of the typical houses in conditioned mode

- Openable window area

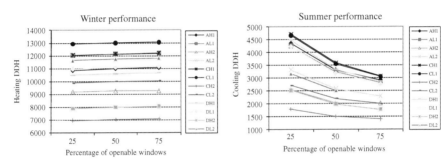

Fig. 6.19 Projected effect of openable window area on the seasonal performance of the typical houses in free running operation mode

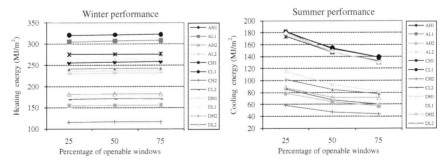

Fig. 6.20 Projected effect of openable window area on the seasonal performance of the typical houses in conditioned mode

- Window to wall ratio (North and South orientation)

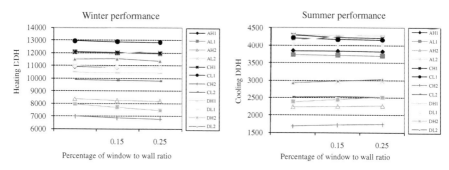

Fig. 6.21 Projected effect of window to wall ratio in north and south orientation on the seasonal performance of the typical houses in free running operation mode

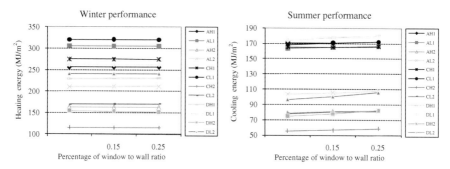

Fig. 6.22 Projected effect of window to wall ratio in north and south orientation on the seasonal performance of the typical houses in conditioned mode

- Window to wall ratio (East and West orientation)

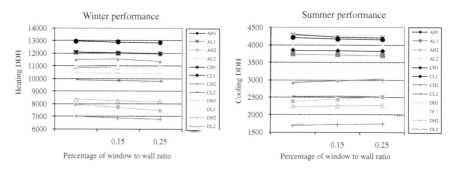

Fig. 6.23 Projected effect of window to wall ratio in east and west orientation on the seasonal performance of the typical houses in free running operation mode

6.1 The Effect of House Envelope Parameters on the Seasonal Performance   139

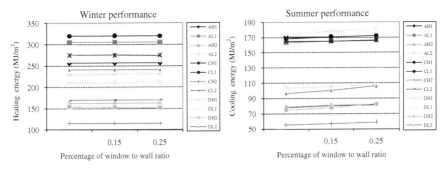

Fig. 6.24 Projected effect of window to wall ratio in east and west orientation on the seasonal performance of the typical houses in conditioned mode

- Internal wall

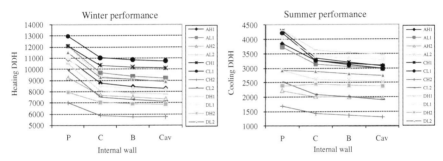

Fig. 6.25 Projected effect of internal wall on the seasonal performance of the typical houses in free running operation mode

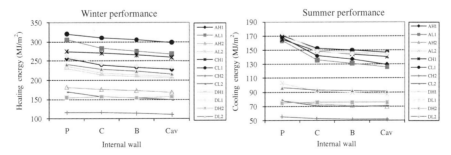

Fig. 6.26 Projected effect of internal wall on the seasonal performance of the typical houses in conditioned mode

- Infiltration

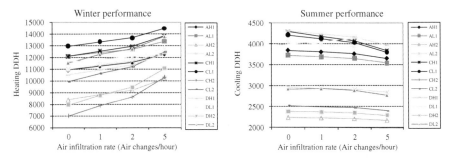

Fig. 6.27 Projected effect of infiltration on the seasonal performance of the typical houses in free running operation mode

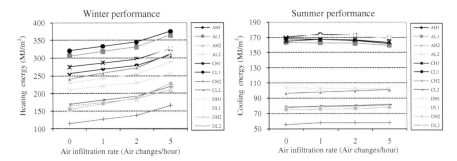

Fig. 6.28 Projected effect of infiltration on the seasonal performance of the typical houses in conditioned mode

Index

A
ABS (Australian Bureau Statistics), 71–72
Absolute value, 77
Absorbance, 75–76, 85, 102
Acclimatization, 37, 43, 47
Accuracy of rating systems, 23
 accuracy of HERS, 23–24
Accurate house rating system, 117
AccuRate software, 22, 63–64, 68–69, 71, 73, 85, 94
Achievement of sustainability, 19
 See also Sustainability
Acoustic comfort, 14
ACTHERS, 14, 58
Active system design, 24
Actual energy consumption, 15
Actual energy cost, 23
Actual energy performance, 22
Actual performance, 19, 54, 64
Adaptive comfort models, 40–43, 67
Adaptive comfort standard, 41–42
Adaptive thermal comfort models, 39–41
Aggregated score, 124
Aggregation technique, 128
Air
 speed, 34, 68, 70
 velocity, 32–33, 36
 ventilation, 41, 69, 76, 85
Air conditioning, 2, 46, 76
Algorithm, 11, 35, 55, 73, 122–123, 126
 simplified algorithm, 73
Annual energy requirements, 21, 24, 26, 43, 54, 78–79, 81–82, 84, 91, 93, 108–109, 119, 121, 126
Annual thermal performance, 78–100, 102, 108–110
Appliances, 9–10, 12, 17, 20
ASHRAE standard, 33, 35–36, 41, 69
Assessing building energy efficiency, 24–25
Australia, 1, 3, 8, 13–14, 18–19, 22, 47, 55, 57–66, 69, 71–72, 108–109, 125
Australian Climate, 13–14, 58, 64–66

B
Benchmark tool, 59
BEP (Billed Energy Protocol), 16
BERS, 14, 58
BESTTEST, 9, 64
Beta coefficients, 111
BREDEM, 11
Brundtland Report, 1
Bubble marker, 125
Building
 envelopes, 10, 13, 21, 74–75, 87, 107, 111
 operation mode, 118
 orientation, 75, 88
Building Energy Rating Schemes, 19–24
Building energy tools, 9
Building performance assessment, 31, 53–54
Bureau of Statistics, 71

C
Calculation, 2, 7–9, 12, 15, 17, 54–55
Canada, 8–10, 13, 18
Canadian homes rating system, 10
Ceiling insulation, 76, 78–79, 102, 110–112
Certification, 8, 10–11, 13, 15
Cessation times, 16
CHEENATH, 14, 58, 60
CHEETAH, 14, 58
Circumstantial restriction, 43
Climate
 cold, 11, 64–65
 hot-dry, 64
 humid, 13, 33, 41, 47, 64, 68
 moderate, 3, 13, 18–20, 23, 31, 33, 47, 65, 67–68, 109, 122

severe, 23, 65
warm, 38, 84
Climate responsive, 111
Climatic
  data, 65, 67
  parameters, 66, 109
Cloud cover, 66
Combined system, 42, 125–126
Comfortable indoor condition, 18, 25, 39
Comfort condition boundaries, 56, 67
Comfort index
  See also Indices
Comfort performance, 103
Comfort temperature, 35, 39–40, 43–45, 67–69, 84
Conditioned floor area, 17, 20
Conditioned mode, 3, 21, 45, 47, 54, 73, 78–82, 84–103, 105, 110–112, 125, 127–128
Conflict in HERS, 3–4
Constant criteria, 56
Cost-effective improvements, 18
Cost effectiveness, 14, 18, 54
Criteria for modeling, 56–74
Criteria of thermal comfort, 34, 41
'criterion- referenced' measurment, 118
Cultural features, 40
Customized limit, 118

D

DDH (Degree Discomfort Hours)
  annual degree discomfort hours, 45, 47, 78–79, 81, 91, 93, 101, 119–121
  cooling degree hours, 86, 88
  cooling' discomfort hours, 45
  heating degree hours, 85, 87–88, 94
  heating' discomfort hours, 45
DDH with area weighting, 74, 78
DDH without area weighting, 74, 78
Degree hours, 43–45, 78, 85–88, 94
  overheating degree hours, 85
Denmark, 11–12
DER (Dwelling Emission Rate), 11
Design
  features, 53, 74, 100–103, 108–112, 126
  paradigm, 19
  quality, 117, 126
Diagnostics, 25
Diffuse irradiance, 66
Disabled people, 42
Dry bulb temperature, 33, 35, 40, 65–66, 68
Dwellings, 5, 7, 11, 13–14, 16, 22, 24, 31, 37, 44, 47, 56, 77–100
  See also Residential buildings

E

Ecological criteria, 19
EEM (Energy-efficient mortgages), 18
Efficiency categories, *see* Score bands
Efficiency standards, 11
Efficient architectural design, 47, 65, 75, 86–87, 90, 122, 128–129
Embodied energy, 16, 19, 54
ENE-RATE, 16
Energy
  based rating schemes, 18, 23, 47, 53, 74, 100, 117
  behaviour, 16
  bill, 18, 23
  conservation, 2, 13, 24
  consumption, 1–4, 7–21, 23–25, 31, 40, 53, 59, 64, 70, 77, 111, 122, 124, 128
  cooling, 14, 33, 45, 58, 60, 74, 81, 84–85, 88, 93–94, 96, 98–99
  cost, 17–19, 22–23
  crisis, 1, 10
  demand, 1, 2, 4, 16, 20, 24
  efficient building, 2, 10, 20, 31, 46, 53
  efficient design, 2, 13, 20, 24, 53, 125
  efficient development, 18
  -efficient houses, 18
  indicator, 20
  inventory, 12
  normalization, 16
  performance, 4, 10–11, 18, 22, 46–47, 54, 103, 117–118, 121
  performance index, 117–118
    See also Indices
  plus, 64
  related fixed components, 17
  simulation, 55
  supply, 1, 3
  total supplied, 16
Energy requirement
  cooling, 33, 45, 58, 60, 74, 81, 84–85, 93–94, 96, 98–99
  heating, 11, 45, 84, 86, 94
Enter procedure, 109
Environmental impact, 14, 18–19, 54
Environmental issues, 18
Environmental parameters, 32, 35
Environmental performance, 9, 16, 64
Environmental temperature, 34, 64, 68, 70
EPBD (Energy Performance of Building Directive), 11
ET*(Effective Temperature), 40
EUROCLASS, 16
Europe, 8, 10–13, 16

Index

EUROTARGET, 16
Evaporation, 33, 65
Experimentation, 54
External walls, 63, 75, 79–81, 84–86, 89, 97

F

Fabric building variables, 108
Fanger, 17, 32, 34–36, 38, 43–44, 47
Financing, 2, 18
FirstRate, 14, 58, 108
Five Star Design Rating, 13
Fixed limit option, 118
Floor area, 12, 17, 20, 63, 74, 82, 101
Floor insulation, 76, 82–84, 102, 110, 112
Framework
    new, 4, 5, 16, 26
    proposed, 125–126
    prototype rating, 125
    regulatory, 3, 112, 125
Free running mode, 23, 45–46, 54, 69, 74, 78–101, 103, 111, 121, 125, 127–128
Fuel, 10, 12, 15, 17
Fuzzy logic, 16

G

GBA (Government Buildings Agency), 44
Glazing type, 75–76, 90–91, 102, 110, 112
Grading on a curve, 118

H

Heating and cooling condition, 69–70
HERS (House Energy Rating Scheme), 2–4, 7–15, 18–26, 55, 58, 64–65, 69, 71, 122–123, 125–126, 128
HFRS (House Free running Rating Scheme), 25, 26, 65, 69, 122–123, 125–126, 128
Home buyers, 7, 111
House construction
    HW house (Heavyweight houses), 79–83, 86–88, 90–92, 94, 97, 100
    LW houses (Lightweight houses), 26, 72, 81, 83–84, 86–87, 90, 92, 94, 97, 100, 108
House envelope, 77
House operation mode, 43, 74, 77, 86, 89, 90–91, 93, 99–101, 111
    See also Conditioned mode; Free running mode
House type
    DS houses (Double Storey houses), 20, 26, 79–80, 82–84, 86–91, 93–94, 96, 98–101, 105–106, 117, 119–121, 125–128
    SS houses (Single Storey houses), 20, 26, 79–81, 83–84, 86, 88–91, 93–94, 96, 98, 100–101, 105–107, 111, 125, 127–128, 117, 119–117, 121
Human heat balance, 41
Humidity
    absolute, 33, 66, 70
    relative, 33, 36, 40, 68–70

I

"Ideal" thermal comfort, 32
IEA (International Energy Agency), 1, 2, 12
Inability of rating systems, 19
Index of Energy Disposition, 15–16
Indicators, 5, 17, 19–21, 26, 31–32, 34, 45–47, 54, 56, 72–74, 95, 101, 103, 105–107, 109–111, 122, 126
Indices, 15–16, 35
Indoor air quality, 14
Indoor climate, 21, 37, 43, 74–75
Indoor comfort condition, 19
Indoor temperature, 24, 37, 43, 64, 67–69, 74, 76, 83, 87, 94
Infiltration, 75–76, 98–100, 102, 110, 112, 140
Inhabitants' behaviour, 21
Internal walls, 63, 74, 76, 97–98, 102, 110, 112
International issues, 1
ISO 7730, 35–36, 44
ISSO, 43–44
ISSO 2004, 43
ITC (Index of Thermal Charge), 16

J

Kyoto Protocol, 1

K

Labelling, 8, 11
    See also Certification
Laboratory based index, 37
    See also Indices
LEED, 9–10
Lighting, 10–12, 17
Linear relationship, 45, 103, 106, 126

L

Marketing, 15, 18
Measurements, 7, 10, 12, 25, 118
MEC, 8
MEP (Monitored Energy Protocol), 16
Method
    bin, 54
    equipment adjustment factor, 15

the modified loads, 15
the normalized modified loads, 15
the original, 15
simplified, 24, 45
Methodologies, 11–12, 14–17
Metrics, 17, 25, 56
"Mixed" comfort, 32
Modelling, 14, 53–112
  process, 77
Mortgage, 2, 7, 9, 18
MRT (Mean Radiant Temperature), 33–34, 36
Multi-criteria, 14, 16–17
  for a building assessment, 16
  method, 14
Multivariate regression analysis, 108–111

M
NatHERS, 14, 24, 58–60, 63–64, 71–72, 118, 125
National regulations, 20
Naturally ventilated houses, 44, 47, 110
Naturally ventilated premises, 37
Natural ventilation, 42, 47, 59, 64, 75, 85, 87–88, 93–94, 96
NHER (National Home Energy Rating), 11–12
Normalised energy use, 17, 20
Norm-referenced measurement, 118
NRCan (Efficiency of Natural Resources Canada), 9

N
Occupancy
  factors, 17, 22
  scenarios, 4, 16, 21–22, 26, 56, 63–64, 70–72, 79, 103, 106
  variables, 17
Occupant behaviour, 21–22, 24, 42, 63, 70
Occupants factor, 71
Occupation time, 16–17, 42, 70–72
Occupied zones, 45, 70–72
OEE (Office of Energy Efficiency), 9, 20
Office buildings, 16, 34, 42–44
Openable windows, 76, 93–94, 102, 110, 112
Operative temperature, 38, 41–43
Outdoor climate, 66, 106
Outdoor temperature, 39–42, 64, 70, 76
Overhang depth, 89–90

O
Pakistan, 39–40
Paper-based check-list, 8
Parametric sensitivity analysis, 77–103, 105, 107–108, 110–112
  See also Sensitivity analysis

Pass/fail rating, 13
Passive design, 53
  passive architectural design, 3, 13, 18
Pearson correlation coefficient, 103
  See also Beta coefficients
Personal parameters, 35–36
Perspiration, 33
Perturbation techniques, 77
Physical phenomenon, 20
PMV (Predicted mean vote), 35–38, 43–44
Poor assessment, 54
Post-occupancy, 54
PPD (Predicted Percentage Dissatisfied), 35–36, 44
Predictors of thermal performance, 108
Prerequisite for an efficient house, 125
Prescriptive, 7–8
Provision of thermal comfort, 18–19, 21, 31

Q
QuickRate, 58

R
Ranking, 5, 16, 22, 24, 72, 101, 103, 110, 112
  of design features, 110, 112
Rating buildings, 4–5, 18–19, 117–122
  features, 17
Rating Index, 20–21, 23
  See also Indices
Rating methodologies, 14–17
Rating scheme in
  Australia, 13–14, 72
  Canada, 9–10
  Denmark, 12–13
  Europe, 10–11
  UK, 11–12
  US, 8–9
Rating system
  current rating schemes, 4, 7, 19, 26, 45, 72, 128
  current, 11, 19, 24, 103
  five-star, 14
    See also Five Star Design Rating
  house, 21, 117
  international, 122
  new, 124, 126–128
  A reliable, 24
  Regulations, 7, 10–12, 20, 53, 118
Regulations for free running buildings, 53
Regulatory framework, 3, 112, 125
Relative humidity, *see* Humidity
Reliability of a rating system, 126–128
Renewable energy, 18, 55
Renovation, 54

Index

Residential buildings, 1, 7, 9, 12, 16, 25, 34, 41–45, 67
Retrofit analysis, 55
Roof colour, 76, 86–87, 102, 110–112
Rule-of-thumb, 54

S

SAP (Standard Assessment Procedure), 11–12
SAVE program, 16
Scatter plots, 103, 107, 126
Score bands, 101, 117–119, 126
Seasonal performance, 77, 79–100
    summer performance, 45, 47, 75, 81–82, 84–96, 99
    winter performance, 79, 81–85, 87–93, 95–99
Sensation of thermal comfort, 31–33, 41
Sensitivity analysis, 77–103, 105, 107–108, 110–112
    See also Parametric sensitivity analysis
Simulation programs, 4, 24, 54–64
Simulations, 4, 8, 14, 23–24, 35, 54–64, 66, 69, 71, 73, 76–77, 83, 86, 92, 96, 100, 103, 105, 107–109, 118
Solar absorbance, 75
    See also Absorbance
SOLARCH, 20, 56
Solar gain, 89
SSPC, 40
Stakeholders, 7, 111
Standard deviations, 118, 121
Standardized coefficient, 109, 112, 124
Standards
    building standards, 8, 23
    thermal comfort standards, 35, 37, 40, 42
Suspended floor, 82–84
    See also Floor insulation
Sustainability, 1, 4, 18–19, 28, 55
Sustainable development, 1, 19, 122

T

Theory of educational measurement, 118
Thermal Comfort, 4–5, 7, 13–14, 16–19, 21, 23, 25–26, 31–47, 53–54, 56, 64, 66–70, 84, 102–111
Thermal comfort condition, 17, 54, 56, 66–70, 111
Thermal comfort index, 36–39
Thermal comfort models, 17, 34–37, 39–41, 43

Thermal discomfort, 36
Thermal mass, 88, 97–98
Thermal neutrality, 39–41, 45, 67–69
Thermal performance
    actual performance, 19, 54, 64
    of buildings, 4–5, 20–21, 25, 31, 43, 55–74, 84, 102–111
    of dwellings, 5, 77–100
    of residential building, 43–45
Thermal performance analysis, 100–102
Thermal sensation, 33–38, 40–41, 68, 74
Thermostat settings, 4, 12, 22–23, 56, 68–70
Threshold value, 118
Typical houses, 56–63, 65, 69, 74, 76–103, 105, 108, 110–112, 126

U

United Kingdom (UK), 2, 11–12
United States of America, 8–9

V

Variation of activity, 34
Victorian scheme, 13

W

Wall colour, 76, 84–86, 102, 110, 112
Wall insulation, 76, 79–82, 86, 102, 110–112
Water supply, 10, 11, 17
Weather data, 58, 66
Weighted exceedance hours, 72
Weighting factor, 44
Window covering, 75–76, 92–93, 102, 112
Window to wall ratio, 56, 76, 93–96, 102, 110, 112
Wind speed, 58, 66
Winning houses, 19

Z

Zero energy buildings, 53
    See also Energy, efficient building
Zone
    bed, 22, 42, 71–72, 74, 79–81, 86–87, 89, 94, 101, 106
    conditioned, 42, 59, 69–74, 97
    living, 22, 67–68, 71–73, 79–81, 83–84, 86–89, 94, 101, 106
    sleeping, 68
    unconditioned, 97